SpringerBriefs in Food, Health, and Nutrition

Springer Briefs in Food, Health, and Nutrition present concise summaries of cutting edge research and practical applications across a wide range of topics related to the field of food science.

Editor-in-Chief
Richard W. Hartel
University of Wisconsin—Madison, USA

Associate Editors
J. Peter Clark, *Consultant to the Process Industries, USA*
John W. Finley, *Louisiana State University, USA*
David Rodriguez-Lazaro, *ITACyL, Spain*
David Topping, *CSIRO, Australia*

T0214354

For further volumes:
http://www.springer.com/series/10203

Lucy J. Robertson

Cryptosporidium as a Foodborne Pathogen

 Springer

Lucy J. Robertson
Institute for Food Safety
    & Infection Biology
Norwegian School of Veterinary Science
Oslo, Norway

ISSN 2197-571X   ISSN 2197-5728 (electronic)
ISBN 978-1-4614-9377-8   ISBN 978-1-4614-9378-5 (eBook)
DOI 10.1007/978-1-4614-9378-5
Springer New York Heidelberg Dordrecht London

Library of Congress Control Number: 2013953876

© Lucy J. Robertson 2014
This work is subject to copyright. All rights are reserved by the Publisher, whether the whole or part of the material is concerned, specifically the rights of translation, reprinting, reuse of illustrations, recitation, broadcasting, reproduction on microfilms or in any other physical way, and transmission or information storage and retrieval, electronic adaptation, computer software, or by similar or dissimilar methodology now known or hereafter developed. Exempted from this legal reservation are brief excerpts in connection with reviews or scholarly analysis or material supplied specifically for the purpose of being entered and executed on a computer system, for exclusive use by the purchaser of the work. Duplication of this publication or parts thereof is permitted only under the provisions of the Copyright Law of the Publisher's location, in its current version, and permission for use must always be obtained from Springer. Permissions for use may be obtained through RightsLink at the Copyright Clearance Center. Violations are liable to prosecution under the respective Copyright Law.
The use of general descriptive names, registered names, trademarks, service marks, etc. in this publication does not imply, even in the absence of a specific statement, that such names are exempt from the relevant protective laws and regulations and therefore free for general use.
While the advice and information in this book are believed to be true and accurate at the date of publication, neither the authors nor the editors nor the publisher can accept any legal responsibility for any errors or omissions that may be made. The publisher makes no warranty, express or implied, with respect to the material contained herein.

Printed on acid-free paper

Springer is part of Springer Science+Business Media (www.springer.com)

Contents

Chapter 1
Introduction to *Cryptosporidium*: The Parasite and the Disease

Cryptosporidium is a genus of protozoan parasites within the phylum Apicomplexa. This phylum, traditionally considered to consist of four clearly defined groups (Adl et al. 2005), the coccidians, the gregarines, the haemosporidians and the piroplasmids, contains thousands of species (Adl et al. 2007), all of which are parasitic, and include some very important disease-causing organisms. These include haemosporidians such as *Plasmodium* species that cause malaria and piroplasmids such as the various *Babesia* species, while included among the coccidians are the multitude of *Eimeria* species, which are of enormous veterinary importance, and also *Toxoplasma gondii* that is of both veterinary and public health importance.

The actual placing of the *Cryptosporidium* genus within the Apicomplexa family tree has been controversial, even after sequencing of the genome of some species; although traditionally placed with the coccidians, and with various characteristics in common with this class, phylogenetic evidence suggests that it is probably more closely related to the gregarines (Leander et al. 2003). With the currently rather tenuous taxonomic framework for the Apicomplexa (Morrison 2009), it may be safest to state that the current placing of *Cryptosporidium* within the Apicomplexa phylum awaits resolution; the most recent classification considers *Cryptosporidium* as a separate group within the Apicomplexa (Adl et al. 2012).

To date approximately 25 different species of *Cryptosporidium* have been formally described, as well as various genotypes. Many of the species are relatively host-specific, but some species are somewhat promiscuous in terms of host specificity, and approximately 50 % of the species have some degree of zoonotic potential, indicating that they have the potential to infect humans as well as the more typical animal species. One species of *Cryptosporidium*, *C. hominis*, is almost exclusively associated with human infection, and the majority of human infections are due either to infection with this species or with the least host-specific species, *C. parvum*.

The other major species that have been demonstrated to be important as human pathogens include *C. meleagridis*, which is primarily associated with infections in turkeys but has also been relatively commonly identified in children in South America, and *C. cuniculus*, which has previously been particularly associated with

L.J. Robertson, *Cryptosporidium as a Foodborne Pathogen*, SpringerBriefs in Food, Health, and Nutrition, DOI 10.1007/978-1-4614-9378-5_1, © Lucy J. Robertson 2014

infections in rabbits, but was also the aetiological agent in an outbreak of waterborne cryptosporidiosis. Some species of *Cryptosporidium* have only been associated with sporadic cases of human infection (e.g. *C. ubiquitum*, which is more usually associated with infections in sheep and cervids) or are particularly associated with infections in immunocompromised patients (e.g. *C. suis* infections, which is generally diagnosed in infections in pigs, and *C. felis* infections, more commonly associated with infections in cats). The sporadic nature of some of these infections or the fact that they apparently only cause illness in immunocompromised patients suggests that these species are not well adapted to the human host, but may nevertheless occasionally establish in such hosts.

For the purposes of this Springer Brief, we focus only upon those species of *Cryptosporidium* that are of importance with respect to public health; this includes the human-adapted species, *C. hominis* and the various species or genotypes with proven zoonotic potential (Table 1.1). Other species that may be of importance to veterinary health but are apparently of negligible relevance to human health will not be considered further.

For the species of *Cryptosporidium* predominantly associated with human infection, *C. hominis* and *C. parvum*, molecular identification beyond the species level (subtyping) has become important for both epidemiological reasons (e.g. for identifying potential transmission routes or sources of infection in outbreaks) and also for understanding the phylogeny of these species and attempting to understand their evolution. Various molecular tools have been developed using genetic targets such as microsatellites and minisatellites and also specific genes, in particular the gene coding for a 60 kDa glycoprotein, gp60. In addition to the sequence heterogeneity of this gene making it useful as a *Cryptosporidium* subtyping target, being one of the most polymorphic markers identified in the *Cryptosporidium* genome, it is also of biological relevance as the protein for which it codes is located on the surface of apical region of invasive stages of the parasite (Xiao 2010). Thus, sequence information at this gene provides a biological possibility of associating parasite characteristics, including clinical presentation, with subtype family. Nevertheless, as a single locus, the resolution of the gp60 gene is relatively low compared with a multi-locus approach, and the development of a standardised multi-locus fragment size-based typing (MLFT) scheme that could be integrated with epidemiological analyses would improve some public health investigations (Robinson and Chalmers 2012).

Cryptosporidium spp. are generally considered to have a global distribution; human infections with *Cryptosporidium* have been reported from more than 100 countries (Fayer 2008). Although prevalence estimates vary between countries, between age groups and between study types, it is clear that children in developing countries are most affected (Shirley et al. 2012); cryptosporidiosis has been suggested to be responsible for around 20 % of cases of childhood diarrhoea in developing countries (Mosier and Oberst 2000), and in 2004, the common link of cryptosporidiosis with poverty encouraged its inclusion in the WHO 'Neglected Diseases Initiative' (Savioli et al. 2006).

Within Europe, several countries (>25) collect data on cases of cryptosporidiosis and report them to the European Center for Disease Control (ECDC). Although the

Table 1.1 Overview of different species and genotypes of *Cryptosporidium* that have been associated with human infection[a]

Species of *Cryptosporidium*	Primary host	Overview of some notable features of human infection	Relevant references[b]
C. andersoni	Cattle	Individual case reports only (e.g. from Malawi); rarely zoonotic, infection may be associated with compromised host nutritional status	Morse et al. (2007)
C. baileyi	Chickens and turkeys	Very few cases reported; associated with immunocompromised status, and reported as disseminated (not confined to intestinal tract)	Ditrich et al. (1991)
C. bovis	Cattle	Very few human cases reported	Khan et al. (2010)
C. canis	Dogs	Sporadic cases diagnosed in both immunocompetent and immunocompromised patients, largely in tropical countries. Asymptomatic or associated with diarrhoea	Gatei et al. (2002, 2008), Cama et al. (2003, 2007, 2008)
C. cuniculus	Rabbits	First identified as pathogenic to humans during a waterborne outbreak, but sporadic cases also diagnosed	Chalmers et al. (2011), Robinson and Chalmers (2010)
C. fayeri	Marsupials	Single human case report only	Ryan and Power (2012)
C. felis	Cats	Sporadic cases diagnosed in both immunocompetent and immunocompromised. More commonly diagnosed in humans than *C. canis* and with a wider geographic distribution, including European countries	Insulander et al. (2013), Cieloszyk et al. (2012), Chalmers et al. (2009), Gatei et al. (2008), Cama et al. (2007, 2008), Raccurt (2007)
C. hominis	Humans	Most common cause of human cryptosporidiosis in many parts of the world, including USA, Australia and Africa. Some sporadic animal infections reported	Xiao (2010)
C. meleagridis	Turkeys, chickens, humans	Cases in both immunocompetent and immunocompromised patients reported. Particularly associated with infection in Peru, with this infection as prevalent as *C. parvum* in some studies	Insulander et al. (2013), Silverlås et al. (2012), Xiao (2010), Chalmers et al. (2009), Cama et al. (2003, 2007)
C. muris	Rodents	More common in immunocompromised patients, but sporadic cases diagnosed in immunocompetent patients also	Al-Brikan et al. (2008), Muthusamy et al. (2006), Palmer et al. (2003)

(continued)

Table 1.1 (continued)

Species of *Cryptosporidium*	Primary host	Overview of some notable features of human infection	Relevant references[b]
C. parvum	Common in preweaned ruminants, especially calves	Very common. In some study areas the most common cause of human cryptosporidiosis	Xiao (2010)
C. scrofarum	Pigs	Individual case reports only	Kvác et al. (2009)
C. suis	Pigs	A few reports from immunocompromised patients	Wang et al. (2013), Cama et al. (2007)
C. tyzzeri	Rodents	Individual case reports only	Rasková et al. (2013)
C. ubiquitum	Sheep and cervids in particular	Sporadic cases frequently diagnosed in both immunocompetent and immunocompromised patients	Fayer et al. (2010), Chalmers et al. (2011), Elwin et al. (2012a), Cieloszyk et al. (2012)
C. viatorum	Humans; no other hosts identified to date	Sporadic cases particularly associated with European travellers to tropical countries	Insulander et al. (2013), Elwin et al. (2012b)
C. wrairi	Guinea pigs	Occasional sporadic cases reported	Azami et al. (2007)
Cryptosporidium chipmunk genotype	Rodents	Occasional sporadic cases reported	Insulander et al. (2013)
Cryptosporidium horse genotype	Horses	Sporadic cases only	Chalmers et al. (2009), Elwin et al. (2012a)
Cryptosporidium mink genotype	Mink	Sporadic cases only	Ng-Hublin et al. (2013)
Cryptosporidium monkey genotype	Monkeys	Sporadic cases only	Elwin et al. (2012a)
Cryptosporidium skunk genotype	Skunk	Sporadic cases only	Chalmers et al. (2009), Elwin et al. (2012a)

[a]Information in the table adapted and updated from Robertson and Fayer (2012)
[b]Not intended as an overview of all case reports

actual incidence is almost certainly higher than that reported and reporting rates are likely to be influenced by factors other than the actual occurrence of infection (e.g. laboratory capabilities, medical awareness of the disease and other national policies and idiosyncrasies), the accumulated data do provide a useful background for comparison. According to the annual epidemiological report from 2012 (ECDC 2013), between 2006 and 2010 the overall incidence of reported cases of cryptosporidiosis has remained approximately stable, ranging from 2.29 cases per 100,000 persons in 2010 to 2.74 cases per 100,000 persons in 2009. The highest rate of confirmed cases was reported by the UK (7.37 per 100,000), followed by Ireland (6.58 per 100,000) and Sweden (4.20 per 100,000); four countries (Cyprus, Estonia, Poland, Slovakia) reported zero cases, five countries reported only one or two cases each (Bulgaria, Czech Republic, Lithuania, Luxembourg, Malta) and nine countries did not notify *Cryptosporidium* cases to ECDC. These differences in incidence are likely to reflect differences in national surveillance systems and diagnostic practices rather than an actual picture of the distribution of *Cryptosporidium* infection in Europe. However, meaningful information can be drawn from the age distribution data, with the highest confirmed case rate reported in the age group 0–4 years (12.19 per 100,000).

As many of the human infections with *Cryptosporidium* are probably derived from animals, either directly or indirectly, it would be useful to have meaningful prevalence data on infections in animals. Although several studies regarding the prevalence/incidence of *Cryptosporidium* infection in individual animal species and regions have been published, there are wide regional, population-specific and age-specific variations, as well as huge variations in study design regarding how the prevalence data have been collected. Thus, in considering cryptosporidiosis as a foodborne infection, relevant data must be sought for each study and, although broad statements can be made (such as that cryptosporidiosis is particularly prevalent in preweaned calves), it is not possible to generalise without oversimplification.

For all *Cryptosporidium* species, the life cycle is direct (no intermediate host) but nevertheless rather complex, containing both a sexual cycle and an asexual cycle. The transmission stage is the oocyst; oocysts are the only exogenous stage in the *Cryptosporidium* life cycle and are excreted in the faeces of infected hosts. These small oocysts (although there is interspecies variation, for most species—including *C. parvum*, *C. hominis*, *C. cuniculus*, *C. ubiquitum*, *C. viatorum* and *C. meleagridis*—the oocysts are approximately spherical, possibly slightly ovoid and around 3–5.5 μm in diameter) have a tough wall of 3 or 4 layers, with an inner layer of glycoprotein that appears to provide strength and flexibility (Jenkins et al. 2010). The average thickness of the walls is around 50 nm, and they contain about 7.5 % total protein, approximately 2 % hexose (glucose, galactose, mannose, talose, glucofuranose, D-glucopyranose and D-mannopyranose) and also medium- and long-chain fatty acids and aliphatic hydrocarbons (Jenkins et al. 2010). Although the dityrosine bonds that are considered to stabilise *Eimeria* oocyst walls appear not to occur in *Cryptosporidium* oocysts, disulphide bonds in the *Cryptosporidium* oocyst wall protein (COWP) have been suggested to contribute to wall strength by forming an extensive matrix (Robertson and Gjerde 2007).

Each oocyst contains four naked sporozoites, and these are developed and fully infectious upon excretion (unlike for other genera of Apicomplexa such as *Eimeria* spp., *Cystoisospora* spp. or *Toxoplasma gondii* for which a period of development in the environment post-excretion is necessary). Furthermore, unlike these other Apicomplexan species, the sporozoites are naked, rather than being contained within sporocysts within the oocyst. Infection with *Cryptosporidium* is initiated when a viable oocyst is ingested by a susceptible host. This may be direct faecal-oral ingestion, or via a vehicle such as contaminated water or food. The infective dose is, theoretically, a single oocyst; however, not all species of *Cryptosporidium* are equally infectious to all people, not all strains or subtypes of a species are equally infective to all people and not all people are equally susceptible to the same strain, subtype or species. Infectious dose studies have been conducted for *C. hominis* (Chappell et al. 2006), *C. parvum* (Chappell et al. 1999; DuPont et al. 1995; Okhuysen et al. 1999, 2002) and *C. meleagridis* (Chappell et al. 2011). These have all shown similar results, with successful infection established in healthy, immunocompetent adults, with resulting symptomatic infection. Infectious doses are varied, but tend to be in the 10–100 range according to study. For example, the median infectious dose for *C. hominis* was found to be between 10 and 83 oocysts (Chappell et al. 2006), but another study demonstrated that the infective dose for *C. parvum* ranged from below 10 to over 1,000, depending on isolate (Okhuysen et al. 1999). Dose–response analyses should not, however, be confined to considerations of the parasite as host variations are also likely to have an impact (Teunis et al. 2002a, b).

When viable oocysts are ingested by an appropriate susceptible host, they usually excyst in the small intestine (*C. baileyi* can be associated with intra-tracheal infection in chickens) where the resultant sporozoites invade epithelial cells—and locate epicellularly (within the cell but not within the cytoplasm)—and they develop to the trophozoite stage. Repeat cycles of asexual reproduction result in the production of huge quantities of meronts, which divide to form merozoites, and each mature merozoite leaves the meront to infect another host cell, with the accompanying destruction of the initial host cell. This asexual cycle results in enormous multiplication of the parasites. A sexual cycle, gamogony, involves the development of microgamonts and macrogamonts, with microgametes produced from the microgamonts fertilising the macrogamont, and ultimately resulting in oocyst production. The oocysts sporulate while within the host and may excyst within the same host, resulting in reinvasion of the epithelial cells and continuation of the infection. Alternatively the oocysts are excreted in the faeces and are immediately infectious to the next host. It has been suggested that those oocysts that reinvade the same host prior to excretion have thinner walls than those that are excreted. Due to the asexual cycle, which results in multiplication of the parasite, thousands of oocysts are excreted from an infected host. A daily excretion rate of over 10^9 oocysts has been reported from AIDS patients with symptomatic cryptosporidiosis (Goodgame et al. 1993), while in healthy, immunocompetent adults experimentally infected with *C. parvum* oocysts, excretion varied with dose, with total oocyst excretion ranging from around 5.5×10^4 to around 8×10^8 (Chappell et al. 1996). Interestingly, in the latter study, higher doses produced fewer oocysts, such that the 'yield' of oocysts was below 1 for doses of 100,000 and a million oocysts and was highest for the dose of 500 oocysts.

In the human host, cryptosporidiosis is usually an enteric disease. It is generally characterised by watery diarrhoea—often voluminous and sometimes mucoid, but rarely containing blood—abdominal pain, nausea, vomiting and related symptoms. The diarrhoea is most usually described as acute, but can also be persistent. In some individuals, however, infection may be largely asymptomatic. Symptoms other than diarrhoea have been reported, and while the spectrum of symptoms depends to some extent on the host (age, immunity, nutritional status), parasite factors are also important, including the species and the number ingested. Putative factors for virulence have been proposed and tend to be factors involved in aspects of host-pathogen interactions, from adhesion and locomotion to invasion and proliferation (Bouzíd et al. 2013). This approach is built upon use of both immunological and molecular techniques, but information can also be obtained in the clinical setting. For example, one study suggests that non-intestinal sequelae (joint pain, eye pain, headache) are associated only with *C. hominis* infection, not with *C. parvum* infection (Hunter et al. 2004a). However, comparison of symptoms reported in patients with cryptosporidiosis in the Milwaukee outbreak (Mac Kenzie et al. 1994) for which *C. hominis* was identified as the aetiological agent (Sulaiman et al. 1998), with symptoms in patients with cryptosporidiosis in patients infected during a small outbreak where *C. parvum* was the aetiological agent (Rimšelienė et al. 2011), demonstrated that while abdominal/intestinal symptoms were rather similar, for the outbreak in which *C. parvum* was the aetiological agent, the symptom 'sore throat' was reported more frequently (39 % patients in the *C. parvum* outbreak compared with 17 % of patients in the Milwaukee outbreak). However, patient factors should also be taken into account, and as the *C. parvum* outbreak was largely restricted to schoolchildren, this may have affected symptoms and symptom perceptions. Nevertheless, and regardless of parasite species or subtype, immunocompromised individuals and children in developing countries are most affected by cryptosporidiosis, and the relationship of infection with growth faltering, malnutrition and diarrhoeal mortality is in need of further exploration (Shirley et al. 2012). It should be noted that *Cryptosporidium* has been identified as being one of the four main aetiological agents that are associated with serious childhood diarrhoea in developing countries (Kotloff et al. 2013).

While cryptosporidiosis is usually self-limiting in immunocompetent individuals, a high relapse rate has been reported; in one study, 40 % of patients, all of whom were immunocompetent, reported the recurrence of intestinal symptoms after resolution of the acute stage of illness, and this was not affected by whether infection was with *C. hominis* or *C. parvum* (Hunter et al. 2004a).

In immunocompromised patients the symptoms are often more severe, and infection may become chronic, debilitating and potentially life-threatening, with high volumes of diarrhoea, potential for the spreading of infection beyond the primary site, and severe weight loss.

It is worth noting that in animal *Cryptosporidium* infections, the symptoms appear to depend highly on parasite adaptation to the host and host age/immunological status, although results from different studies vary, and for many species infections are largely asymptomatic. Nevertheless, it has been repeatedly observed that infections of some domestic animals, particularly calves, with some species of *Cryptosporidium*, particularly *C. parvum*, may result in severe infection, usually with acute diarrhoea as the main symptom. In some animals, especially young animals and particularly in association with concomitant infections or conditions, this may even be fatal.

Symptoms of cryptosporidiosis commonly start about 1 week after infection, but incubation periods from 24 h to as much as 2 weeks have also been reported.

For example, in experimental human infection studies with *C. hominis*, the period between infection and symptoms ranged between 2 and 10 days (Chappell et al. 2006). In this study the mean duration of diarrhoea was around 6 days, ranging from 2 to 10 days. For the *C. meleagridis* experimental human infection study (Chappell et al. 2011), the period between infection and symptoms was very similar to that for *C. hominis* (4–7 days), but duration of diarrhoea tended to be shorter (2–4 days); however, differences in experimental setup mean that these studies are not directly comparable.

As cryptosporidiosis is predominant in children, some studies have investigated whether *Cryptosporidium* infection in children in developing countries may result in long-term health consequences, particularly with respect to cognitive function and failure to thrive. A study from Brazil indicated that early childhood cryptosporidiosis was associated with reduced fitness at 6–9 year of age, even when controlling for current nutritional status (Guerrant et al. 1999), and a study in Peru demonstrated that *C. parvum* infection in children has a lasting adverse effect on linear (height) growth, especially when the infection is acquired during infancy and when the children are nutritionally stunted prior to infection (Checkley et al. 1998). Furthermore, early childhood diarrhoea (not necessarily due to cryptosporidiosis) has been associated with long-term cognitive deficits (Niehaus et al. 2002) and impaired performance at school (Lorntz et al. 2006).

Innate immunity seems to be of importance in the control of *Cryptosporidium* infection, in particular that the infected epithelium seems to provide a trigger when infection occurs, activating intracellular signalling cascades, shaping local inflammatory responses and directly inactivating parasites (McDonald 2008). It is possible that the unusual extracytoplasmic but intracellular location of *Cryptosporidium* provides partial protection from immunological attack. Although adaptive immunity is accepted as being necessary for establishing immunity against *Cryptosporidium*, investigations using mouse and in vitro infection models also indicate that innate immune responses have a key role in the development of resistance. Mechanisms that have been shown to be of partial importance include Toll-like receptors, inflammatory molecules and antimicrobial peptides. The role of complement, however, is unclear. Natural killer (NK) cells as the source of proinflammatory cytokines, particularly interferon gamma (IFN-γ), appear to be of importance, although results obtained in vitro have not always been supported by in vivo models, and the cytotoxic activity of NK cells may also have a role. In contrast with many other intracellular pathogens, CD4$^+$ T-cells appear to play an important role in the immune response against this parasite, and this has been shown experimentally using replacement and depletion studies in immunocompromised mice.

Although infection with *Cryptosporidium* results in the production of parasite-specific antibodies in both the serum and intestine, studies using B-cell-deficient mice indicate that antibodies are not essential for combating the infection. However, there is evidence that passive acquisition of maternal antibodies with colostrum reduces parasite reproductive capacity, and thereby symptoms. However, contrasting results have been obtained from different studies, and it seems probable that while antibodies have a deleterious effect on *Cryptosporidium* infection, they are not a vital factor in host resistance.

Diagnosis of cryptosporidiosis usually depends on the demonstration of oocysts (or, less commonly, their antigens or DNA) in faecal samples. Although antibody-based detection in serum/plasma, saliva or faeces is also possible for demonstrating exposure to *Cryptosporidium*, it is only of proper diagnostic benefit if seroconversion can be demonstrated, as otherwise a positive result can indicate past exposure, current infection or both.

For faecal samples, a concentration technique such as formol-ether (ethyl acetate) or flotation (often sucrose or sodium chloride) is usually used prior to microscopy. As the oocysts are very small, the use of a staining technique, particularly using antibodies labelled with a fluorochrome and screening with fluorescence microscopy (immunofluorescent antibody testing; IFAT), is recommended. IFAT is considered to be a gold standard, although other techniques such as modified Ziehl–Neelsen (mZN) or auramine phenol staining may also be used successfully. However, some oocyst staining methods do not perform well on oocysts that have been preserved in polyvinyl alcohol fixatives. Furthermore, in a number of survey studies in which mZN has been used for identification, the lack of sensitivity and specificity of the method has resulted in overestimation of the prevalence of infection (Chang'a et al. 2011).

Antigen tests, such as ELISA-based assays, have also been developed, and rapid tests based on the same principle (immunochromatographic assays) are also commonly used. The disadvantage with such rapid assays is not only are they relatively expensive but also may be of low sensitivity if oocyst numbers are low (Johnston et al. 2003; Robertson et al. 2006), while low specificity has been suggested may be due to interpretation of weak lines of colloidal carbon as potential (weak) positives (Johnston et al. 2003; Robertson et al. 2006). However, the tests are very simple to use and provide results within minutes at the point of care and have therefore gained popularity among some users. Nevertheless, even when relatively good sensitivity and specificity are reported, it is suggested that these tests might be a useful addition to, but not a substitute for, microscopy-based methods (Weitzel et al. 2006).

It is generally accepted that screening for intestinal parasites, including *Cryptosporidium*, by PCR will become increasingly common in the decades ahead and, as the feasibility improves due to automation and high-throughput facilities, might even replace microscopy of faecal concentrates (Stensvold et al. 2011). The specific detection of parasite DNA in stool samples using real-time PCR is particularly likely to become a method of choice, especially a multiplex approach allowing simultaneous testing for a range of different pathogens. However, microscopy of faecal concentrates currently remains a cornerstone, not only because many diagnostic laboratories do not have the technological capabilities for PCR, but also because of some limitations in extracting DNA from faecal material; when formalin has been used as a storage medium or for formol-ether sedimentation, then this inhibits PCR, but also faecal samples themselves can include a range of PCR inhibitors, including bilirubin, bile salts and complex polysaccharides. Furthermore, as the specificity of primers and probes means that the only sequences detected will be those for which they were designed—then an unusual species may not be identified if the primers/probes selected are too specific. Care is important in selecting primers and loci, and it should always be considered whether the intention is to amplify genus-specific DNA or species-specific DNA.

Despite the necessity of using molecular technologies for determining species, the utility of microscopy should not be undervalued; during the acute stage of *Cryptosporidium* infection when oocyst excretion is high, the use of microscopy for reaching a positive diagnosis can be vastly faster (and cheaper) than using molecular techniques.

Chemotherapy of cryptosporidiosis is perhaps the area that seems to have lagged behind the most regarding our knowledge of this infection, and successful treatments for cryptosporidiosis—in both humans and animals—remain elusive. Those medications that are used successfully for the treatment of other coccidian parasites appear to be ill-suited towards combating cryptosporidiosis. The reason for the failure of anticoccidial preparations to eradicate *Cryptosporidium* infections has been attributed to the unique taxonomic placement of this parasite; although initially considered to be a coccidian, later studies suggest that it should be in a separate group. When *Cryptosporidium* was first identified as being of particular importance in immunosuppressed patients, particularly those with AIDS, a vast array of different chemotherapeutic, immunomodulatory and palliative agents were tried in that population. This has resulted in a list of around 100 agents that are ineffective or are not consistently effective. Nevertheless, one treatment (nitazoxanide) has been FDA-approved for symptom alleviation in immunocompetent humans and has shown promise for treating cryptosporidiosis in animals.

For immunocompromised patients with HIV infection, the development of effective highly active antiretroviral therapies (HAART) has been of greater value for decreasing mortality due to cryptosporidiosis than any parasite-targeted treatment, and this has perhaps reduced the urgency for developing an effective treatment. In developing countries, however, antiretroviral therapy coverage is often limited and thus cryptosporidiosis in HIV patients, particularly in association with other insults to health, may prove fatal.

Furthermore, the identification of *Cryptosporidium* as one of the four main aetiological agents associated with serious childhood diarrhoea in developing countries (Kotloff et al. 2013) indicates that there is still a need to develop an effective chemotherapy targeting this parasite and that is efficacious in the most vulnerable populations.

Chapter 2
Transmission Routes and Factors That Lend Themselves to Foodborne Transmission

Transmission of *Cryptosporidium* infection occurs when an appropriate number of infectious *Cryptosporidium* oocysts are ingested by a susceptible host. Transmission can be hand-to-mouth and may be associated with unhygienic conditions or high-risk behaviour. Although sporadic cases of cryptosporidiosis in the community can be of individual clinical significance, particularly if the infected person is immuno-compromised, the major public health importance of *Cryptosporidium* lies in the potential for outbreaks to occur when drinking water, recreational water or food become contaminated with infectious *Cryptosporidium* oocysts. Such contamination can result in several individuals becoming infected via the same transmission vehicle, and, for drinking water in particular, this can be of considerable community and economic importance, with tens, hundreds or even thousands of people at risk of infection (Clancy and Hargy 2008). Additionally, when a large-scale outbreak occurs, with many infections occurring simultaneously in a particular community due to contamination of a common vehicle, then, due to the excretion of yet more oocysts into the environment, the potential for subsequent environmental contamination increases accordingly and thus the potential for secondary spread.

Particular factors in the biology of *Cryptosporidium* mean that this parasite is particularly suited to foodborne or waterborne transmission. These are:

- The large numbers of infective oocysts that are excreted by an infected individual into the environment (calves infected with *C. parvum* may produce as many as 6×10^7 oocysts per gram of faeces, and a single infected calf may excrete 4×10^{10} oocysts during its second week of life and 6×10^{11} oocysts during its first month of life; Uga et al. 2000; Nydam et al. 2001)
- The relatively low infectious dose
- The robustness of the oocyst and its ability to survive in the environment; experimental results suggesting that oocyst viability is retained for at least a month in damp conditions and in the absence of freeze-thaw cycles (Robertson et al. 1992; Robertson and Gjerde 2006) and that oocysts are to some extent resistant to commonly used disinfectants such as chlorine (King and Monis 2007)

- The relatively small size of the oocysts (3–5 µm in diameter) that enables penetration of sand filters used in the water industry
- The possibility for zoonotic transmission for some species of *Cryptosporidium*— this means that there is greater for potential for environmental spread and contamination
- The possibility for onward contamination or transfer by transport hosts such as insects; promiscuous-landing synanthropic flies have been particularly associated with the carriage of protozoan parasites to food (Conn et al. 2007)

Taken together, not only do these factors mean that there is a high potential for possible vehicles of infection such as food or water with *Cryptosporidium* oocysts but also that they will probably survive on such vehicles in sufficient quantities to pose an infection risk to susceptible hosts.

It is worth noting that in a recent risk ranking of foodborne parasites (http://www.fao.org/food/food-safety-quality/a-zindex/foodborne-parasites/en/ and http://www.who.int/foodsafety/micro/jemra/meetings/sep12/en/; see also Robertson et al. 2013), *Cryptosporidium* was ranked as number 5 out of 24 potentially foodborne parasites in terms of importance as a foodborne pathogen, exceeded only importance by *Taenia solium*, *Echinococcus granulosus* and *Echinococcus multilocularis* and *Toxoplasma gondii*. This relatively high ranking (compared with, e.g. *Giardia duodenalis* that was ranked in position 11) reflects not only our increasing awareness of cases and outbreaks of foodborne cryptosporidiosis, but also the lack of an effective treatment and the importance of cryptosporidiosis as a significant cause of morbidity and mortality, particularly in children in developing countries (Kotloff et al. 2013). In this risk-ranking exercise, fresh produce, fruit juice and milk are listed as the food commodities that are most likely to act as transmission vehicles for *Cryptosporidium*, with these choices based on the documented outbreaks of cryptosporidiosis recorded in the literature (Robertson et al. 2013).

Chapter 3
Documented Foodborne Outbreaks of Cryptosporidiosis

Waterborne transmission of cryptosporidiosis is well known, and outbreaks of waterborne cryptosporidiosis have been extensively documented. Indeed, as *Cryptosporidium* oocysts that contaminate water are probably more likely to remain infectious for longer than oocysts that contaminate food products (as *Cryptosporidium* oocysts survive best under moist, cool conditions), waterborne transmission has a higher potential to result in infection, and in a larger number of people, than food-borne transmission. Therefore, as the most dramatic outbreaks of cryptosporidiosis have been waterborne, research interest and funding has been particularly directed towards this transmission route. Waterborne outbreaks of cryptosporidiosis have been extensively reviewed; of 325 outbreaks of human disease attributed to the waterborne transmission of pathogenic protozoa (from the beginning of records up until around 2003), the majority of them (approximately 51 %) were caused by *Cryptosporidium* infection (Karanis et al. 2007). Although most of these are suggested to be *C. parvum* infections, the lack of molecular characterisation methods at the time of many of these outbreaks, coupled with the fact that the majority of different species of oocysts are morphologically indistinguishable, means that it is likely that a large proportion of these outbreaks were actually due to *C. hominis* infections (and possibly some other species of *Cryptosporidium*). A follow-up review of more recent outbreaks (Baldursson and Karanis 2011) indicated that between 2004 and 2010 *Cryptosporidium* spp. continued to be the dominant aetiological agent of waterborne outbreaks of protozoan disease, with more than 60 % of the 199 documented outbreaks due to *Cryptosporidium* infection. Most outbreaks of waterborne cryptosporidiosis are reported from developed countries—indeed the largest outbreak of waterborne cryptosporidiosis recorded to date occurred in Milwaukee, USA, in 1993 (Mac Kenzie et al. 1994), and over 400,000 people were estimated to have acquired symptomatic infection. Nevertheless, common sense tells us that waterborne cryptosporidiosis is probably more likely to occur in less developed countries, where the infection is perhaps more likely to be endemic and where those infrastructures that are necessary for ensuring a safe drinking water supply, such as an intact sewage disposal system, effective catchment control measures and efficient

L.J. Robertson, *Cryptosporidium as a Foodborne Pathogen*, SpringerBriefs in Food, Health, and Nutrition, DOI 10.1007/978-1-4614-9378-5_3, © Lucy J. Robertson 2014

water treatment, may be suboptimal. The fact that outbreaks or cases of waterborne cryptosporidiosis are rarely reported from less developed countries (in the review by Karanis et al. (2007) no waterborne outbreaks of cryptosporidiosis from developing countries are recorded; in the review by Baldursson and Karanis (2011), just one (from Malaysia is noted) is probably more related to the endemicity of cryptosporidiosis and other diarrhoeal infections in these countries, which makes detection of outbreaks and vehicles of transmission difficult. The lack of detection and monitoring systems, both at the public health level and at the water treatment level, also mean that such outbreaks are less likely to be recognised. A waterborne outbreak would probably be less likely to be detected in a community where a considerable proportion of the population are already infected with *Cryptosporidium*, because, unless the outbreak cases were particularly distinctive, the outbreak cases would be unlikely to show up against the background of non-outbreak cases.

That waterborne cryptosporidiosis is of greater public health significance than foodborne cryptosporidiosis is also reflected in the fact that standard methods for analysis of water for *Cryptosporidium* oocysts were first developed between 5 and 15 years ago (e.g. US EPA Methods 1622 and 1623; ISO Method 15553), while as of today, there is no widely accepted standard method for investigating food products for these parasites. Nevertheless, the value of monitoring of drinking water (either post or pretreatment) for *Cryptosporidium* contamination has been the subject of considerable debate, as the methods are both expensive and time consuming, and interpretation of data can be difficult. However, while it is widely agreed that general performance indicators (e.g. turbidity, particle removal, pressure in distribution system) are probably of most importance for ensuring the microbial safety of the drinking water supply, it is also acknowledged that regulatory, event-driven monitoring of source water for contamination, using a site-specific monitoring programme, may provide important data that can be used as data input for risk assessments for an individual water source and thereby enable the application of appropriate barriers. Analysis of water samples for *Cryptosporidium* oocysts also supplies critical information in the event of an outbreak, and molecular analysis to the subtype level of oocysts detected in water supplies indicates the origins of contamination, enabling effective controls to be implemented.

Although outbreaks of waterborne cryptosporidiosis are more like to occur and be recognised than outbreaks of foodborne cryptosporidiosis, and although less people are likely to be infected in outbreaks of foodborne cryptosporidiosis, nevertheless, such events do occur and several have been documented (Table 3.1).

Despite there being many similarities in the potential for outbreaks of foodborne cryptosporidiosis and the potential for outbreaks of foodborne giardiasis, there are important differences. Although many more outbreaks of foodborne cryptosporidiosis have been documented than outbreaks of foodborne giardiasis (see the companion Springer Brief "*Giardia* as a Foodborne Pathogen"; Robertson 2013), the vehicles of infection for *Giardia* seem to be much broader than for *Cryptosporidium*. Thus, for the nine outbreaks of foodborne giardiasis tabulated, nine different associated food matrices are listed, while for the 21 outbreaks of foodborne cryptosporidiosis listed in Table 3.1, three different associated food matrices predominate: salad

Table 3.1 Documented outbreaks of foodborne cryptosporidiosis from 1990 to date, including probable vehicle of infection[a]

No. of cases (age group)	Country (year of outbreak)	Species and subtype of *Cryptosporidium* (when known)	Associated food and suspected source of contamination	Other information	Reference
Over 300, mostly adults	UK (2012)	*C. parvum, gp60* subtype[b] information not provided	Ready-to-eat, precut mixed salad leaves from a major supermarket chain. Source of contamination not identified	Weaker evidence suggests that smaller amounts of contaminated salad, possibly spinach, were distributed through another supermarket chain	Anonymous (2013)
16 cases, all adults	Sweden (2010)	*C. parvum, gp60* subtype[b] IIdA20G1e identified in 2 cases	Unknown. Source of contamination not identified, but an isolate of the same subtype has previously been identified in a Swedish calf, suggesting a zoonotic connection	These two outbreaks occurred almost simultaneously, but were separated geographically and epidemiologically and were only identified from investigation of cryptosporidiosis cases in a 3rd city in Sweden	Gherasim et al. (2012)
Probably around 90 cases, all adults	Sweden (2010)	*C. parvum, gp60* subtype[b] IIdA24G1 identified in 6 cases	Salad garnish on chanterelle sauce. Source of contamination not identified, but not thought to be a food handler		
46 (Adults and children at a youth summer camp)	USA (2009)	*C. parvum, gp60* subtype[b] IIaA17G2R1 identified in 7 cases and most animals	Sandwich bar ingredients (particularly ham and lettuce, possibly tomatoes or onions) possibly contaminated from livestock at the camp farm (calves, goat kids, piglets)	Difficulties in the outbreak investigation limit the strength of the findings, but a food source of infection seems definitive	Collier et al. (2011)

(continued)

Table 3.1 (continued)

No. of cases (age group)	Country (year of outbreak)	Species and subtype of *Cryptosporidium* (when known)	Associated food and suspected source of contamination	Other information	Reference
74 (63 children and 11 adults at a mountain farm/wildlife park camp)	Norway (2009)	*C. parvum*, gp60 subtype[b] IIa A19G1R1 identified in cases. Another outbreak occurred 2 years later at the same time, in which this same subtype was identified in both patients, and animals (lambs and goat kids) and direct infection was suspected	Possibly peeled carrots contaminated via a food handler with oocysts either derived from human infection or from farm livestock (lambs, goat kids, calves)	Foodborne transmission postulated, but other transmission routes also possible (e.g. contaminated water, direct transmission associated with animal handling)	Rimšelienė et al. (2011)
18	Sweden (2008)	*C. parvum* of four gp60 subtypes[b] identified: IIaA16G1R1 (*n*=10), IIaA15G2R1 (*n*=1), IIdA22G1 (*n*=3), and IId19G1 (*n*=1)	Arugula salad. Source of contamination not identified	Suspected source of infection was based on a case–control study. Molecular results from 15 cases	Insulander et al. (2013)
72 (Adults)	Finland (2008)	*C. parvum* identified by PCR at the COWP gene in 1 sample. No subtyping conducted	Lettuce mixture, packed in Sweden and originating from 5 different European countries. Source of contamination not identified	12 cases lab confirmed. No oocysts detected on salad remains	Pönka et al. (2009)

No. infected	Location (year)	Organism/method	Food/source	Confirmation	Reference
21 (Adults)	Sweden (2008)	*C. parvum* identified by PCR at the COWP gene in 13 samples. No subtyping conducted	Béarnaise sauce containing freshly chopped parsley (imported in plastic bags from Italy) and added shortly before serving (after heating). Source of contamination not identified, but not thought to be a food handler	16 cases lab confirmed	Insulander et al. (2008)
4 (Adults)	Japan (2006)	*C. parvum gp60* subtype[b] IIa A15G2R1 confirmed for 3 of the cases	Yukke—Korean-style beef tartar and/or raw liver. Source of contamination not identified	All cases lab confirmed	Yoshida et al. (2007)
Two outbreaks reported—but numbers infected not provided	Germany (2006)	No information provided	Milk and dairy products. No information on potential source of contamination provided	Insufficient information to assess whether these were confirmed foodborne outbreaks or whether the foodborne transmission route is speculative	EFSA (2007)
99 (Adults)	Denmark (2005)	*C. hominis* identified	Salad bar items (particularly whole carrots, grated carrots, red peppers). Source of contamination unknown, but infected food handler (although not someone involved in food preparation) speculated as source of contamination of the buffet	12 cases lab confirmed	Ethelberg et al. (2009)

(continued)

Table 3.1 (continued)

No. of cases (age group)	Country (year of outbreak)	Species and subtype of Cryptosporidium (when known)	Associated food and suspected source of contamination	Other information	Reference
144 (Adults and children)	USA (2003)	C. parvum gp60 subtypes[b] IIaA15G2R1 and IIaA17G2R1 identified in 11 patient samples and C. ubiquitum in 1 patient sample. A sample from a jug of cider found also to contain C. parvum DNA of subtype IIaA17G2R1	Ozonated apple cider. Source of contamination not identified	23 cases lab confirmed. This study suggested that the ozonisation regime used was insufficient to inactivate Cryptosporidium oocysts in the cider	Blackburn et al. (2006)
8 (Children)	Australia (2001)	Species of Cryptosporidium not identified	Unpasteurised milk; tested Cryptosporidium positive by ELISA. Source of contamination not identified	It is illegal to sell unpasteurised cow milk for human consumption in Queensland. This milk was labelled as unpasteurised pet milk	Harper et al. (2002)
152 (College students)	USA (1998)	C. hominis identified in 25 samples from 19 patients	Probably fruits and vegetables associated with a specific meal and contaminated by an infected food handler	Foodborne outbreak in a canteen setting. Descriptive epidemiology indicated foodborne infection route	Quiroz et al. (2000)
31 (Various)	USA (1996)	Species of Cryptosporidium not identified	Apple cider. Water from contaminated well that was used for washing apples	Convincing foodborne outbreak in a community setting	Anonymous (1997)
50 (School children)	UK (1995)	Species of Cryptosporidium not identified	Cow's milk. Source of contamination not identified, but inadequate udder hygiene suggested	Pasteurisation failure at commercial, on-farm dairy supplying a local school	Gelletlie et al. (1997)

15 (Not stated; attendees of a social event)	USA (1995)	Species of *Cryptosporidium* not identified	Chicken salad containing cooked chopped chicken, pasta, peeled and chopped hard-boiled eggs, chopped celery and chopped grapes in a seasoned mayonnaise dressing, prepared in domestic kitchen of a licensed daycare home and speculated to be the origin of contamination, possibly indirectly from a child in her care	As the food handler did not submit a stool sample for analysis, the origin of contamination remained speculative	Anonymous (1996)
160 (Students and staff)	USA (1993)	Species of *Cryptosporidium* not identified	Apples collected from the ground in an orchard grazed by infected calves. Assumed that the contamination originated from the calves	Oocysts detected in apple cider, on the press and in calves from the farm, but lack of PCR analysis means that molecular support is missing	Millard et al. (1994)
At least 13 infants from nursery and orphanage	Russia (1990)	Species of *Cryptosporidium* not identified	Yoghurt type drink (kefir), with on-farm contamination assumed	Person-to-person spread may also be responsible for at least some of the cases. Oocysts detected in milk filters	Romanova et al. (1992)
19 (7 adults, 12 children)	UK (1984)	Species of *Cryptosporidium* not identified	Raw sausage suggested by case–control study, but insufficient evidence to be definitive. Raw milk also considered a possible vehicle	Concurrent outbreak of campylobacteriosis; no oocysts detected in milk samples	Casemore et al. (1986)

[a]Adapted and updated from information previously published in Robertson and Chalmers (2013) and Robertson and Fayer (2012)

[b]*gp60* subtypes are identified by sequencing part of the 60 kDa glycoprotein gene. The nomenclature is derived from Sulaiman et al. (2005)

bar items and salad garnishes, dairy products and apple cider. It can be speculated that one possible reason for this difference is that in nearly all foodborne giardiasis, the contamination event with *Giardia* cysts actually originates from a food handler, and therefore, contamination of a product occurs depending on the contact history of the food handler. In contrast, a substantial amount of foodborne cryptosporidiosis could be of zoonotic origin, for which the contact scope (types of food contaminated due to contact with animal faeces) is likely to be more limited. The likelihood that foodborne cryptosporidiosis is often of zoonotic origin is perhaps borne out by comparing the species of *Cryptosporidium* most associated with waterborne outbreaks with the species of *Cryptosporidium* most associated with foodborne outbreaks. In a review of waterborne cryptosporidiosis, selected outbreaks from England and Wales between 2001 and 2002 in which molecular typing was conducted are tabulated (Chalmers 2012); of the four outbreaks associated with drinking water, three were associated with *C. hominis* and one with *C. cuniculus*. In other major outbreaks of waterborne cryptosporidiosis, such as the Milwaukee outbreak (Mac Kenzie et al. 1994) and a more recent outbreak in Östersund, Sweden, in 2010 in which 27,000 people suffered from waterborne cryptosporidiosis, *C. hominis* has, again, been identified as the aetiological agent (Robertson 2014). In contrast to this apparent predominance of the anthroponotic species of *Cryptosporidium*, *C. hominis*, as the aetiological agent in waterborne outbreaks, for foodborne outbreaks *C. parvum* appears to predominate. Of the 12 outbreaks in which species information was obtained that are listed in Table 3.1, 10 involve zoonotic *C. parvum* (1 outbreak also has a single case of *C. ubiquitum* as well as other cases of *C. parvum*) and only 2 have *C. hominis* as the aetiological agent; for both of these latter outbreaks, a food handler is considered the most probable source of contamination. While the identification of *C. parvum* as the aetiological agent does not necessarily mean that the source of the contamination was an animal rather than a food handler, it does open up for this possibility and seems more likely given the background that many of the waterborne outbreaks are not, apparently, due to *C. parvum*.

Nevertheless, it should be noted that food handlers are considered to represent a common route of contamination of food, including with parasites such as *Cryptosporidium* (Greig et al. 2007), and that any item that is handled by an infected food handler with poor hygiene may act as an infection vehicle for cryptosporidiosis (Girotto et al. 2013). However, although various surveys of food handlers for endoparasites have been conducted in different countries, particularly developing countries, *Cryptosporidium* is rarely reported. This could be because the food handlers included in such surveys are usually adults and, as most people in developing countries will probably have been infected as children, they will therefore have developed immunity. However, it could also be that the infections are not identified, as *Cryptosporidium* is a more challenging protozoan parasite to identify than larger parasites such as *Giardia*. Nevertheless, a study from Venezuela identified *Cryptosporidium* infection in 14 of 119 food handlers (over 10 % prevalence) (Freites et al. 2009). One intriguing study investigated the role of money as a potential environmental vehicle for transmitting parasites among food-related workers in Egypt (Hassan et al. 2011); over 50 % of banknotes and coins were found to be contaminated with parasites, with *Cryptosporidium* one of the most predominant. Thus, handling of money and then food may be one method by which food might be contaminated with *Cryptosporidium* oocysts regardless of the personal hygiene of the food handler.

Another interesting contrast between the foodborne outbreaks of giardiasis and those of cryptosporidiosis is the geographical distribution, with a surprising number of outbreaks of foodborne cryptosporidiosis reported from Nordic countries (in Table 3.1, which includes 21 outbreaks from ten different countries, 7 of the outbreaks are reported from Nordic countries and 4 of these are reported from Sweden). Although USA is also strongly represented for outbreaks of foodborne cryptosporidiosis (six outbreaks included in Table 3.1), this seems more understandable on a population-size basis and also in comparison with waterborne outbreaks and outbreaks of foodborne giardiasis. The potential reasons for the apparent skewed distribution of foodborne cryptosporidiosis towards Nordic countries have been previously reflected upon (Robertson and Chalmers 2013), with possible reasons considered including prolonged survival of oocysts in the Nordic climate, greater exposure of the Nordic population due to elevated consumption of higher-risk products (possibly including imported foods) and better outbreak investigation and reporting. It was concluded that although consumers in these countries seem to be less concerned about microbiological contamination of foods than consumers in other European countries, there was no clear reason for the apparently skewed distribution apart, perhaps, from superior infrastructure and resources in these countries enabling more directed investigation and reporting (Robertson and Chalmers 2013).

As *Cryptosporidium* oocysts are inactivated by desiccation and heat treatment, it is those food substances that are intrinsically moist and are most often consumed raw or very lightly cooked, such as salad vegetables and cold beverages, which would seem to be the most likely vehicles for infection. Furthermore, contamination of vegetables and fruit can, of course, take place at any point along the field to fork continuum. Irrigation water and splash up from the soil are both particularly relevant potential contamination sources for fresh produce, while shellfish, such as oysters, also have the potential to be contaminated in situ before harvesting, and unpasteurised dairy products, especially milk, have the potential to be contaminated from the animal being milked; a case–control study of cryptosporidiosis patients in West Germany demonstrated that drinking unpasteurised milk was significantly associated with cryptosporidiosis (Freidank and Kist 1987). As shellfish in particular, but also fresh produce, fresh-fruit beverages and dairy products are usually kept cool and moist before consumption, then *Cryptosporidium* oocysts probably have similar chances of surviving on such products as they do in a water body. However the number of people likely to be exposed to a contaminated food product is probably considerably less. Additionally, when a solid food product is contaminated, then the contamination may be localised to a particular area or portion of the product, such that not all consumers of the same product, even when derived from the same lot, are necessarily exposed. This potential for localisation of contamination provides a further confounder for epidemiological investigation of potentially foodborne outbreaks.

Interestingly, while various outbreaks of cryptosporidiosis have been associated with consumption of fresh produce, at the same time several investigations have indicated a negative association between cryptosporidiosis and eating raw vegetables. For example, a matched case–control study in USA involving 282 persons with laboratory-identified cryptosporidiosis and 490 age-matched and geographically matched controls demonstrated that eating raw vegetables was protective against cryptosporidiosis (Roy et al. 2004), a case–control study of sporadic cryptosporidiosis in UK with genotyping of isolates from case patients and including 427 patients and 427 controls similarly demonstrated that eating raw vegetables and eating tomatoes were strongly negatively associated with illness due to *C. parvum* infection (Hunter et al. 2004b) and a case–control study investigating risk factors for sporadic cryptosporidiosis in Australia demonstrated that eating raw carrots had a statistically significant protective effect (Robertson et al. 2002). It has been proposed that this apparently protective effect associated with consumption of raw vegetables may be due to repeat exposure to low numbers of *Cryptosporidium* oocysts on raw vegetables, allowing the development of protective immunity (Bouzid et al. 2013). In contrast, however, a study from urban dairy farms in Dagoretti, Nairobi, suggested that consumption of vegetables was a greater source of risk for acquiring zoonotic cryptosporidiosis than consumption of milk (Grace et al. 2012); however, this latter study was based on a theoretical stochastic simulation model rather than a matched case–control study, and therefore, it might be expected that somewhat differing results might be obtained.

It should be noted that while there is a clear risk of *Cryptosporidium* infection associated with consumption of leafy vegetables that may have been contaminated in the field, for root crops the risk seems to be much smaller, unless contaminated by food handlers. Indeed, a study designed to predict the number of *Cryptosporidium* infections in the UK transmitted by consumption of root crops grown on agricultural land to which treated sewage sludge had been applied (according to the regulations and guidance regarding treatment before application and the interval between application and harvest) suggests that the risk is remote, being only one infection in the UK every 45 years (Gale 2005).

Nevertheless, it should be emphasised that not all countries have a system in place for reporting foodborne diseases, and, even in countries where such a system is established, under-reporting is considerable. This is largely due to lack of knowledge of the physician or the victim regarding the possible aetiological role of foods, particularly for parasitic infections such as cryptosporidiosis. Additionally, once a suspected contaminated food product has been eaten or discarded, then it is unavailable for analysis, and thus, confirmation of the foodborne route of infection becomes impossible. In the majority of foodborne outbreaks of cryptosporidiosis described in Table 3.1, *Cryptosporidium* contamination was never identified, or even sought, on the implicated product as none was available for analysis. Indeed, the only outbreaks in which the implicated food items were convincingly demonstrated to be contaminated were two of the apple cider outbreaks in USA (Millard et al. 1994; Blackburn et al. 2006). In one of these (Blackburn et al. 2006) the remaining content of a jug of cider that one of the patients had drunk had been stored in the fridge and was available for examination; PCR analysis demonstrated it to be positive for *Cryptosporidium*. While epidemiological analysis may enable the investigators to pinpoint a particular food item in a large outbreak, for individual cases or a cluster of a small number of cases, this may be impossible. Indeed, even in large outbreaks, it may be difficult to determine a suspect food, particularly with a buffet-type situation involving multiple food types and different combinations of foods being eaten both by infected patients and noninfected people attending the same meal. An outbreak of cryptosporidiosis in 1984 in UK, involving 19 individuals and which also involved a concurrent outbreak of campylobacteriosis, suggested the possibility that consumption of either raw milk (partially prompted by the simultaneous campylobacteriosis outbreak) or raw sausage (identified via a case–control study) might have been the vehicles of infection (Casemore et al. 1986). However, the lack of sufficient evidence, including no oocysts detected in milk samples (although morphologically similar objects were seen), meant that no conclusion was reached regarding the most probable vehicle of infection.

Chapter 4
Approaches to Detecting *Cryptosporidium* Oocysts in Different Food Matrices

While diagnosis of *Cryptosporidium* infection is generally based upon identification of oocysts in faecal samples, it can also be based upon detection of antigens in faecal samples (or, more rarely, detection of antibodies in blood). However, detection of contamination of vehicles of infection with *Cryptosporidium* oocysts, whether water or food, relies entirely on isolating and identifying either the *Cryptosporidium* oocysts themselves, or DNA from the *Cryptosporidium* oocysts, on the contamination vehicle. It should be noted that the concentration of *Cryptosporidium* oocysts in a faecal sample from an infected individual is likely to be considerably higher than the concentration of oocysts on a contaminated potential vehicle of transmission, and thus detection in food or water is likely to be much more difficult. For this reason, the diagnosis of *Cryptosporidium* infection by identifying oocysts in faecal samples and the detection of contamination by identifying *Cryptosporidium* oocysts in food samples are not entirely comparable.

As previously mentioned, Standard Methods for detecting *Cryptosporidium* oocysts in drinking water have been available for many years and are used in both regulatory laboratories and in the research setting. In brief, these methods involve concentrating particles that are approximately the size of *Cryptosporidium* oocysts (or larger) from a relatively large water sample (minimum of 10 L) by filtration (flocculation and sedimentation can also be used, but are used less frequently, or continuous flow centrifugation may be used), eluting these particles from the filter into a smaller volume, concentrating the smaller volume (often by centrifugation), and then isolating the cysts from other material in the concentrate before detection. While immunomagnetic separation (IMS) is most commonly used for isolation, it is expensive, and other cheaper separation techniques, such as density gradient flotation, are sometimes used when a particular method has not been stipulated by regulatory requirements. Flow cytometry (fluorescent-activated cell sorting) has also been considered as a method in which both concentration and detection could be combined. However, although gaining some degree of popularity in laboratories

L.J. Robertson, *Cryptosporidium as a Foodborne Pathogen*, SpringerBriefs in Food, Health, and Nutrition, DOI 10.1007/978-1-4614-9378-5_4, © Lucy J. Robertson 2014

already proficient in the use of flow cytometry for other applications, the disadvantages of this method (large equipment costs, need for specialised training, reliance on parasites being individually suspended, low sensitivity, potential for false negatives and false positives although specificity can be improved by using multi-labels) mean that it has not been widely accepted generally and has not been included in any of the Standard Methods. Detection of the oocysts is usually performed by drying the final post-IMS concentrate of around 50 μL onto a microscope slide and examining it by IFAT. The fluorescent marker on the monoclonal antibody used is usually fluorescein isothiocyanate (FITC), and additionally, a stain for the sporozoite nuclei, usually 4′6 diamindino-2-phenyl indole (DAPI), is usually used for improving detection by providing additional visual markers for identification. However, other detection methodologies are possible, including using molecular-based detection systems.

In principal, the approach to analysing food matrices for contamination with *Cryptosporidium* oocysts is the same as that used for water. However, apart from clear beverages, filtration of a large volume is not possible (filtration of colloidal liquids such as milk is also impractical), and instead some sort of elution procedure must be used for the food item itself. This is likely to mean that a relatively smaller (in terms of portion size) amount of product can be analysed. Furthermore, the approach to elution is likely to be influenced by both the physical and biochemical nature of the product, in order to optimise removal of the parasites into a fluid phase, but at the same time minimise contamination with particulates or other material that may hamper the subsequent steps in the procedure. Variations in both the biochemical and physical characteristics of different matrices, from meat to dairy products, to fruits and vegetables and to shellfish, mean that a common 'one-method-suits-all' approach is unsuitable and a method that is appropriate for one type of food may result in recovery efficiencies being suboptimal in other matrices.

The final step in analysis is detection. For all the food types listed below, and also for the Standard Methods for analysing water, IFAT usually remains the method of choice and is the method stipulated in the ISO Method (ISO/TC 34/ SC 9/WG) currently under development for analysis of fresh leafy green vegetables and berry fruits for *Cryptosporidium* (and *Giardia*). Given the advances in molecular detection systems within recent decades, it may be surprising that IFAT (which is basically a microscopy-based detection system) has not been supplanted by a technique such as real-time PCR or LAMP, particularly as the equipment required for IFAT, a fluorescence microscope, is highly expensive while PCR equipment is becoming more competitively priced and, in addition, provides the opportunity for simultaneous species identification rather than as a downstream analytical step. In the diagnostic lab, multiplex real-time PCR is more often becoming the method of choice for protozoa diagnostics, including *Cryptosporidium* and other parasitic pathogens in stool samples (Stark et al. 2011; Taniuchi et al. 2011). However, in diagnostics, the numbers of a particular pathogen in a sample are expected to be relatively high, whereas in environmental samples not only are the numbers low, but it may be important to detect non-nucleated parasites, and obviously these will not be detected by methods for which nuclear material is necessary. The reason why it may be important to detect non-nucleated *Cryptosporidium* oocysts is because although such oocysts are of themselves of no public health importance, their presence indicates that the material being investigated has been contaminated with *Cryptosporidium* and that another subsample may contain nucleated, viable parasites. Another reason why molecular detection may be unsuitable for use in the analysis of food and environmental samples is because the range of potential inhibitors is variable and not well known, and differences in matrix types may require that PCR conditions are adjusted per matrix. Nevertheless, some research groups are beginning to publish on occurrence of parasites, including *Cryptosporidium*, in food matrices in which PCR (either standard PCR or qPCR) is used for the detection system. For example, an experimental study in which the recovery of *Cryptosporidium parvum* oocysts in a faecal suspension that was intentionally inoculated onto lettuce leaves was investigated, with comparison of microscopy and PCR as detection methodologies, and although only 0–6.5 % of the total numbers of oocysts inoculated were recovered and detected by microscopy, PCR detection was nevertheless less sensitive than microscopy (Ripabelli et al. 2004). In a survey of ready-to-eat packaged greens in Canada, Dixon et al. (2013) used nested PCR as the detection step for different parasites, including *Cryptosporidium*, subsequent to a simple washing step and concentration by centrifugation (without use of IMS for purification). Unfortunately, this research apparently did not include any seeding experiments to determine the limits of detection, and although all samples that were found to be positive by PCR were also examined by IFAT (and it was reported that 23 of the 32 (72 %) PCR-positive samples were also positive by IFAT), the researchers apparently made no effort to determine whether some samples might have been positive by IFAT and not by PCR (by examining PCR-negative samples by IFAT). Another study compared the use of PCR, IFAT and flow cytometry for detecting *Cryptosporidium* oocysts eluted from fresh produce (lettuce and water spinach leaves) and also in irrigation water in Thailand and found that IFAT and flow cytometry provided similar results, although false-positive results due to autofluorescent algae occurred with flow cytometry but that PCR often failed, with only 2 out of 27 samples positive (Keserue et al. 2012). The authors speculate that the high number of negative results obtained with PCR could, for some samples, be due to low numbers of oocysts, but for other samples was probably due to inhibitors. A comparative trial of different DNA extraction kits for *Cryptosporidium* oocysts seeded into washes from raspberries and basil indicated that some kits might give better results (Shields et al. 2013), although the limit of detection was nevertheless relatively high. Thus, although these research reports indicate the possibility for using molecular methods in such surveys, until more comprehensive research comparing detection methodologies is undertaken and successfully adopted by different laboratories, it is probable that IFAT will continue to be the detection method of choice for the immediate future for analysing food products for contamination with *Cryptosporidium* oocysts.

4.1 Fruits and Vegetables

Both common sense and outbreak considerations indicate that fresh fruits and vegetables are food products that have a relatively high likelihood of being vehicles for transmission of *Cryptosporidium* infection, and therefore, these food products have been the focus of method development. In method standardisation most focus has been placed on this food product group (particularly fresh leafy green vegetables and berry fruits) by the relevant ISO Group (ISO/TC 34/SC 9/WG 6), although other food matrices (specifically fruit juice, milk, molluscs and sprouted seeds) have also been considered. However, it was concluded that either the requirement for a Standard Method for analysing these other food matrices for *Cryptosporidium* oocysts was insufficient at the time of consideration (2011) or that the data available in the scientific literature were insufficient to use as a basis for a Standard Method development. Thus, fresh leafy vegetables and berry fruits have been the sole focus for analytical method standardisation at present (registered in the ISO/TC34/SC9 work programme with the number ISO 18744).

4.1.1 Method Development

One of the earliest published methods for analysing fresh produce for *Cryptosporidium* oocysts involved sonication of sub-portions in a detergent solution, with layering on Sheather's fluid for particularly dirty samples, and detection by IFAT (Bier 1991). The recovery efficiencies from seeded cabbage and lettuce samples were rather low (1 %), and this may be partly due to this method being developed before IMS for *Cryptosporidium* became available. The first published methods that investigated the use of IMS for isolation of *Cryptosporidium* oocysts from eluate from experimentally inoculated fruit and vegetables (Robertson and Gjerde 2000) achieved recovery efficiencies of around 40 % for all matrices except for bean sprouts for which recovery efficiency was significantly lower and more variable. Use of the same method, in which the initial step involves elution into a detergent-based buffer (Di Benedetto et al. 2006, 2007), reported recovery efficiencies of over 70 % from leafy vegetables and around 40 % for vegetables such as tomatoes and peppers, indicating that the method performed acceptably in an independent laboratory.

In the first publication to consider the development of a Standard Method for analysis of fresh produce for parasites, specifically *Cryptosporidium* and *Giardia* (Robertson and Gjerde 2001a), a range of parameters that might affect recovery efficiency of the method were investigated, including sample weight and age and the use of different IMS systems, and with particular consideration of 'difficult' samples such as bean sprouts. The long-term intention of this study was to lay a foundation for future development of a Standard Method and to demonstrate that various aspects should be considered in such development and that methods that demonstrate an improvement over previous approaches do not necessarily represent an end point; as techniques develop or different matrices become of importance, with different characteristics, methodologies should have the capacity to develop also. However, logistic considerations mean that once a Standard Method does become established, it may become more difficult to innovate and improve upon that method; alterations in the proscribed technique may have to undergo a range of independent trials to show equivalency, and, unless there is a good incentive for this, the effort and costs may not be worth it for many laboratories. Thus, ongoing review and evaluation of Standard Methods should be part of the remit of a method standardisation team. However, although this may be the intention, again the costs involved may mean that people with the qualifications and experience to review Standard Methods and evaluate the relative merits of alternatives or innovations do not have the motivation to undertake this task.

In the comparative work of Robertson and Gjerde (2001a), sample freshness was found to be a parameter of importance, as with older samples the initial elution step tends to result in greater quantities of interfering material in the eluate, including cellular debris, microflora and excretory products of microflora and biofilms, specifically bacterial exopolysaccharides. These substances may interfere with the subsequent purification (IMS) and detection procedures (Robertson and Gjerde 2001a). Bean sprouts, in particular, were found to produce copious amounts of interfering debris, even when fresh, presumably due to their method of cultivation that also encourages growth of bacteria. However, in an outbreak situation, one can assume that it is probable that any food available for analysis is unlikely to be fresh, and this should be borne in mind by the analyst.

A series of publications by Cook and colleagues in 2006 and 2007 (Cook et al. 2006a, b, 2007) followed on from this earlier work, first by investigating a range of elution methods (stomaching, pulsification, rolling, orbital shaking) and elution media (Cook et al. 2006a). However, before investigating the efficaciousness of the different media at eluting oocysts from lettuce and raspberries by the various methods, their compatibility with the IMS kit that was to be used was first investigated. As three of the media were apparently not compatible with IMS, they were not further investigated. The results obtained regarding compatibility were, in some cases, rather surprising (e.g. IMS from PBS at pH 7.2 only yielded a recovery efficiency of just over 10 %) and also meant that the membrane-based elution solution that had been previously used in the work by Robertson and Gjerde (2000, 2001a, b) was excluded from further investigation by Cook and colleagues. The final optimised method described by Cook et al. 2006a thus included stomaching in 1 M glycine, with concentration of the eluate by centrifugation, IMS for purification and finally detection by IFAT. The use of the 'best' IMS kit is discussed at length by Cook et al. (2006a), and the IMS kit selected for their work is stated to 'outperform' that used in the work of Robertson and Gjerde (2000, 2001b). However, the kit that Cook et al. (2006a) reported as being superior is apparently no longer available, and in the later publication by the same group (Cook et al. 2007), the same kit that Robertson and Gjerde (2000, 2001b) had used and that also can be used for isolating *Giardia* cysts is employed instead; this is the kit that is mentioned in all Standard Methods. Thus, it is clear that caution must be used in defining Standard Methods and method flexibility that allows for products to change, as well as research advances, must be built in if possible.

The optimal pH for the elution solution is also explored at length in the work of Cook et al. (2006a, b, 2007), although it is not always clear how the results are interpreted. According to tabulated data presented, a rather low pH seems to provide the best recovery efficiency from raspberries, but according to the text of that document, it is stated that increasing the pH above neutral results in larger numbers of oocysts being recovered (Cook et al. 2006a). Other research has suggested that the quantity of particles in the water may also affect recovery efficiency in subsequent detection procedures (Chaidez et al. 2007).

With the method of Cook et al. (2006a), a recovery efficiency of approximately 60 % from lettuce and approximately 40 % from raspberries is recorded. The use of the same method (although the pH is not stated, nor the IMS kit used) by eight different laboratories in a validation trial, involving samples of both raspberries and lettuce artificially contaminated at three different levels and also non-contaminated samples, demonstrated an overall recovery efficiency by the different laboratories of approximately 30 % from lettuce and around 44 % for raspberries (Cook et al. 2006b). Thus, although the results for raspberries were highly similar to those obtained by the laboratory in which the method had originally been developed, those from lettuce were approximately 50 % lower (60 % recovery compared with 30 % recovery). The reason for this considerably lower recovery efficiency is not clear but indicates difficulties with the method in different laboratories that did not occur in the lab that first described the method. In addition, several of the participating laboratories reported detection of oocysts on samples that were actually negative (and found to be negative when checked by the distributing lab). This implies that the counts in the positive samples may have actually been overestimates. This publication thus goes to serve as an important reminder of the value of blind trials in establishing a method, and that data from a highly experienced laboratory that has developed the method may not, perhaps, translate to actual results from a less experienced laboratory. The method has also been used in independent laboratories undertaking surveys, and these labs also report relatively low and/or variable recovery efficiencies, for example, 17 % recovery for lettuce (Amorós et al. 2010) and between 4 and 47 % (mean of 24 %) for a variety of vegetables (Rzeżutka et al. 2010). The use of an internal control for estimating recovery efficiency that has also been recommended in this method has been suggested by Rzeżutka et al. (2010) to be unsuitable, as the inclusion of such a control prevents the subsequent use of PCR for determining species or genotype.

4.1.2 Method Standardisation

The purpose with method standardisation is not only to attempt to ensure that comparable methods are used by different laboratories but also that the methods used, provided that they are conducted by competent and appropriately trained personnel, are likely to provide satisfactory, robust recovery efficiencies.

The method being considered for standardised analysis for green leafy vegetables and berry fruit for *Cryptosporidium* and *Giardia* (ISO Draft 18744) is based very closely on the standard water analysis protocol for these parasites (ISO 15553), although with elution into a specified medium as the first step, either through agitation of the produce by shaking or by stomaching using a stomacher. Elution from the food is followed by concentration of the eluate by centrifugation, isolation of the parasites from other debris by IMS and finally identification by IFAT and DAPI staining, as previously described. The development of this method was based not only the water analysis protocol but also upon publications that described use of this method, or variations of this method and with published recovery efficiencies that were considered to be acceptable. Despite some difficulties with the work of Cook et al. (2006a, b), particularly with regard to confusion over pH and suitability of different IMS kits as noted in the previous section, it should also be recognised that this group was the first to recommend elution into glycine, which is considerably more user-friendly than a detergent-based elution solution. Furthermore, an inter-laboratory trial was organised as part of this research, and this is an important procedure for method validation when considering development of a Standard Method.

The two fresh products included in the method being developed for standardisation, fresh leafy green vegetables and berry fruits, present two different challenges when selecting appropriate elution procedures. For leafy vegetables (e.g. lettuce), there is a large surface area that has the potential for contamination, and for some varieties of such vegetables, the leaves are deeply lobed and/or frilly (e.g. oakleaf varieties of lettuce), such that some leaf areas are protected. For other lettuce varieties, such as Mâche (also known as lambs lettuce), rosettes of leaves are held together in nubs of roots, providing pockets for contaminants to gather and not be readily eluted. Additionally, such pockets are also likely to include soil and other debris that may be an impediment in further steps of the analytical procedure.

For berry fruits, the delicacy of the fruit presents the problem rather than the surface area, as vigorous elution procedures are likely to break the fruit themselves, and the resultant fruit tissue fragments in the elution liquid might impede the subsequent concentration and isolation steps. Some fruits and vegetables also have hairy, rather than smooth, skins, and it may be less easy to remove *Cryptosporidium* oocysts from such produce. Indeed, experiments comparing attachment of *Toxoplasma* oocysts to smooth-skinned blueberries and hairy raspberries demonstrated that they were more likely to remain attached to the raspberries (Kniel et al. 2002), while Armon et al. (2002) report on the tendency of *Cryptosporidium* oocysts to be difficult to remove from courgette (zucchini) surfaces. Thus, not only are such produce perhaps more likely to be contaminated at consumption (parasites not removed by standard household washing), but it will be more difficult to elute the parasites from such produce for detection and identification.

The quantity of sample analysed is a matter for consideration in a Standard Method; for water samples, the volume analysed is considerably over a portion size (minimum of 10 L), but for food samples, it is probably not possible to analyse in an equivalent fashion, and it has been demonstrated that the greater the sample size, then the less efficient the method at recovering parasites (Robertson and Gjerde

2001a). This reduced efficiency is presumably a reflection both of compromised elution efficiency, together with the increased quantity of other debris from the produce in the eluate and that may inhibit or hinder other steps in the analysis process. Thus sample size should be a compromise that is selected to maximise recovery efficiency and also to enable detection of low-level contamination, at least at the infective dose per portion size. This is likely to vary according to the type of sample being analysed, but currently recommended sample sizes according to the ISO 18744 Method are between 25 and 100 g.

Thus, the Standard Method that is under consideration is based broadly on the previous work, particularly that of Cook et al. (2006a, b, 2007), and consists of the three previously described steps of elution, concentration and purification and detection. In the proposed Standard Method, elution is into a glycine buffer, either using a homogeniser (paddle blender) for leafy vegetables (buffer pH at 5.5) or by agitation in a lidded container for more fragile berry fruit (buffer pH at 3.5), as described by Cook et al. 2006a. Concentration and isolation involves centrifugation of the eluate followed by IMS. Detection is using IFAT with DAPI and Nomarski (differential interference contrast) optics to aid in identification. Although the use of an internal control is not an obligatory part of the method, it is included in a note to the method.

4.1.3   Further Research on Method Development

Further research beyond that already achieved leading up to the development of a Standard Method has tended to focus upon aspects of molecular detection (e.g. Shields et al. 2013; Yang et al. 2013). However, other research has explored the use of alternative approaches. A 'proof of concept' study investigated the feasibility of using adhesive tape to remove *Cryptosporidium* oocysts from the surfaces of selected fresh produce for subsequent detection by either IFAT or PCR (Fayer et al. 2013). This approach was found to be successful down to levels of ten oocysts applied per area of fresh produce surface, although at the lower levels of contamination IFAT as a detection method from the tape was found to be more sensitive than PCR, but was also very much dependent on the surface of the produce. The study used apples, peaches, cucumbers and tomatoes, and peaches were found to be particularly problematic to work with due to the hair-like projections (trichomes) on the skin (exocarp) of peaches, providing a physical barrier that prevented the adhesive tape from reaching all the oocysts, particularly those located at the base of the trichomes (Fayer et al. 2013). Thus, as with other methods for analysing fresh produce for contamination with *Cryptosporidium* oocysts, factors in the structure of the produce are important in determining the efficacy of the technique. For example, although the adhesive tape method may be of utility for produce with a low surface area to weight ratio and where the surface is smooth, for produce such as leafy greens where the surface area to weight ratio is large, then a washing method is probably more effective for detecting low numbers of oocysts that may be widely distributed. However, it should be noted that research that investigated removal of

Cryptosporidium oocysts from the surfaces of apples to which they had been experimentally applied found that complete removal of oocysts was not possible; the method used most similar to that described in the Standard Method (agitation in 1 M glycine (pH 5.5) for 15 min using an orbital shaker) did not remove all the oocysts from the apple surfaces, and the most efficient removal (37.5 %) was achieved by rigorous manual washing in water with a detergent and by agitation in an orbital shaker with Tris-sodium dodecyl sulphate buffer (Macarisin et al. 2010b). Not only were some oocysts attached in deep natural crevices in the apple exocarp, but some oocysts appeared to be closely associated with what appeared to be an amorphous substance with which they might have been attached to the apple surface (Macarisin et al. 2010b).

4.2 Shellfish

Although there have been no documented outbreaks of cryptosporidiosis associated with bivalve molluscan shellfish, this product group is recognised as having potential as a vehicle for transmission. Not only are such shellfish traditionally consumed raw or lightly cooked (and it has been experimentally demonstrated that *Cryptosporidium* oocysts survive the process of being steamed in mussels; Gómez-Couso et al. 2006b), but they are also likely to come into contact with parasite transmission stages in sewage outflow or runoff from land due to their preferred locations (intertidal or estuarine areas or areas close to the coast). Pathogens in such waters may become accumulated in bivalve molluscan shellfish tissues due to their particular method of alimentation that involves filtration of large volumes of water and concentration of particles (Robertson 2007).

Although, a widely accepted, optimised method for analysis of shellfish for contamination with *Cryptosporidium* oocysts has yet to be described (Robertson 2007), and currently there is no Standard Method available, some research groups have attempted to develop an optimised method by artificially contaminating shellfish and comparing recovery efficiencies of different methods and approaches of analysis. Different research groups have sometimes reported rather different efficacies of very similar methods. In general, the methods start with a tissue homogenisation step. Although some research group have used gill washings or haemolymph (obtained by drilling a hole in the shell and aspirating the abductor muscle), most researchers apparently agree that tissue homogenates provide better results than gill samples (homogenate or washings) or haemolymph (MacRae et al. 2005; Robertson and Fayer 2012). However, it should be noted that loss of oocysts may be considerable in preparing the tissue homogenate, especially if it is sieved before analysis (Schets et al. 2013), and another research group found that they obtained the best results for analysis of oysters for *Cryptosporidium* oocysts when the haemolymph was kept separate during the homogenisation of the whole oyster meat but was then added to the pellet following diethyl ether extraction of the homogenate (Downey and Graczyk 2007). Preparation of the sample is usually followed by a concentration procedure (frequently centrifugation) and then a purification/isolation procedure, which may be nonspecific (e.g. flotation on caesium chloride or sucrose gradients or lipid extraction) or specific (IMS). Although IMS has been considered useful by the majority of researchers, its utility is affected by the nature of the matrix; for example, one research group found that it performed less effectively with mussels than oysters as the latter were less mucoid (MacRae et al. 2005), while another research group was unable to obtain satisfactory results even with oysters and consequently did not use it for an initial survey (Schets et al. 2007), although employed it in a later survey with some minor adjustments (Schets et al. 2013).

However, the very different biochemical nature of shellfish compared with water concentrates or washings from fruits and vegetables may suggest that a completely different elution approach may be more suitable. Based on the relatively high protein content of shellfish (8–20 % depending on shellfish species), Robertson and Gjerde (2008) developed a pepsin-digestion method that was based on the methodology usually used for the detection of *Trichinella* spp. larvae in meat or for recovering *Ostertagia ostertagi* larvae from the abomasal mucosa of cattle. This method was found to result in relatively high recovery efficiencies (70–80 %) when followed by oocyst isolation by IMS and detection by IFAT (Robertson and Gjerde 2008). This method has since been modified by Willis et al. (2012), with the protein digestion followed by concentration by centrifugation and washing in detergent solution, in order to avoid the expensive IMS procedure. This modified method has the advantage of being considerably cheaper and apparently results in very little reduction in oocyst recovery efficiency. However, for some samples analysed by the modified method of Willis et al. (2012), a large pellet size precluded complete analysis, and this problem might perhaps have been resolved by the inclusion of a further purification step, not necessarily based upon IMS (e.g. flotation). Further comparative research with protein digestion may provide a method that may be considered suitable for standardisation when it has been validated in other laboratories.

Although in the majority of studies of *Cryptosporidium* contamination of fresh produce (fruit and vegetables), IFAT has been shown to be the detection method of choice; several research groups analysing shellfish for *Cryptosporidium* oocysts have used other techniques or combined IFAT with other techniques such as fluorescent in situ hybridisation (FISH) and PCR. Some research groups report PCR as being less sensitive than IFAT, but several authors consider that the different techniques should be used to complement each other, and the limitations of both should be acknowledged. It should be noted that molecular methods may provide the opportunity for simultaneous identification of species, but such species-level determination can also be used downstream from IFAT detection. It is also of relevance to note that the recovery efficiencies using the same method might vary with species of shellfish, with particularly mucoid shellfish, being more likely to have lower recovery efficiencies. One study that compared methods for detecting *Cryptosporidium* oocysts in shellfish reported that the most sensitive method for the detection of *C. parvum* in oocyst-exposed mussels was IMS concentration with IFAT detection (Miller et al. 2006). Molecular methods were also investigated in this study, and although IFAT detection provided superior sensitivity, the authors note that TaqMan PCR provided the possibility for automated testing, high throughput and semi-quantitative results and was therefore superior to conventional PCR.

4.3 Meat

Apart from the outbreak of cryptosporidiosis associated with 'yukke' (Korean-style beef tartar) and/or raw liver (see Table 1; Yoshida et al. 2007), there is a lack of documented outbreaks or individual cases associated with ingestion of contaminated meat or meat products. As a result relatively little research has been directed towards detecting *Cryptosporidium* oocyst contamination of meat, although one method has been published for isolating and detecting *Cryptosporidium* oocysts from beef carcass surfaces (Moriarty et al. 2004). This method, in which fat beef tissue and lean beef tissue are considered separately, contains four steps: (a) elution of the oocysts from the beef using a pulsifier into a suspension medium containing phosphate buffer saline solution with 0.1 % Tween 80 (PBST); (b) concentration of the oocysts in the PBST by membrane filtration; (c) elution of the oocysts from the membrane into a smaller volume (10 mL) of PBST by scraping and vortexing then concentration by centrifugation; and (d) detection using IFAT. The authors report high recovery efficiencies, being over 85 % for fat tissue and over 128 % for lean tissue. That the number of recovered oocysts exceeded the size of the initial inoculum is worrying and suggests either problems in preparation of the inocula initially (a possibility suggested by the authors), or problems in identifying the oocysts subsequently (misidentification of other objects reacting specifically or nonspecifically with the monoclonal antibody); as DAPI was not used as an identification aid in this study and nor was Nomarski/DIC microscopy, the latter possibility seems relatively likely, particularly as nonspecific cross-reactivity of the detecting monoclonal antibody with fat globules from meat has been reported in at least one other study (Robertson and Huang 2012). Another reason for the surprisingly high recovery efficiencies described by Moriarty et al. (2004) could be that the initial oocyst inocula were given relatively little time to bind to the food matrix (allowed to stand for only 15 min between inoculation and recovery); however, this would not result in more oocysts being detected than were originally seeded onto the sample.

An extensive outbreak of waterborne cryptosporidiosis in Sweden in 2010 in which cured meat products were potentially exposed to the contaminated water during their preparation in the factory in the affected town prompted the development of a method for analysis of cured meats for contamination with *Cryptosporidium* oocysts (Robertson and Huang 2012). In the optimised method developed in this study, for which a recovery efficiency of over 60 % was reported from seeded samples, surface sections were removed from the meat and soaked in a detergent solution for 30 min to soften the product. The meat sections and buffer were then stomached, the eluate collected and then a further stomaching was conducted, this time in 1 M glycine. Concentration was by centrifugation, with 1 % deoxycholate added to particularly fatty samples as a dispersant. IMS was used for purification of oocysts from the concentrate, and detection was by IFAT. However, for samples that were particularly fatty, background fluorescence from small fat globules hampered IFAT detection, as clusters of globules resembled oocysts. Addition of 0.1 % Evans Blue (a well-known quenching agent that works by shifting the autofluorescent spectrum to longer wavelengths) to the sample at the same time as the monoclonal antibody reduced this effect (Robertson and Huang 2012).

A study from India investigated goat meat samples for contamination with *Cryptosporidium* oocysts using three detection methodologies (IFAT, Ziehl–Neelsen carbol fuchsin stain and PCR) subsequent to elution and sucrose flotation (Rai et al. 2008). However, the sample preparation description is difficult to follow, and recovery efficiencies of the method are not provided (Rai et al. 2008). In another unpublished study, *Cryptosporidium* contamination has been reported to have been detected in various raw meats (chicken breasts, minced beef, pork chops) from retail outlets using PCR and IFAT for detection (Dixon 2009). However, the actual process used for detection is not supplied, and, again, nor is the recovery efficiency.

4.4 Beverages

There have been several outbreaks of cryptosporidiosis associated with beverages, specifically apple cider and milk, and thus there has been some research directed towards developing methods for analysing beverages for contamination with *Cryptosporidium* oocysts. This is in contrast with *Giardia*, for which, perhaps surprisingly, there have been no outbreaks or individual cases for which a beverage (other than water) has been the vehicle of infection, and thus, there has been little development of methods for analysing beverages for these parasites. It is possible that those methods described here for analysing beverages for *Cryptosporidium* contamination could be adapted for *Giardia* also.

The first investigation of methods for analysing beverages for *Cryptosporidium* contamination was probably that of Deng and Cliver (2000) in which they compared formalin-ethyl acetate sedimentation or sucrose flotation for concentration of oocysts from apple cider and detection of oocysts in the concentrate by acid-fast staining, IFAT and PCR. Sucrose flotation was found to be more efficient than sedimentation in recovering oocysts, and IFAT was found to be the most sensitive detection technique; the authors speculate that inhibitory substances in the apple juice abrogated the efficiency of the PCR (Deng and Cliver 2000). Of the methods attempted in this study, the highest sensitivity achieved was detection of between 10 and 30 oocysts per 100 mL of apple juice, using an IMS method after flotation, and with IFAT as the detection methodology. Another study demonstrated that using a microbead IMS system followed by detection by PCR could detect as few as ten oocysts in 100 mL apple juice (Deng et al. 2000). Nevertheless, in an outbreak of apple cider-related cryptosporidiosis in which contamination was actually detected in the implicated product, the method used was direct centrifugation for concentration, followed by molecular detection (Blackburn et al. 2006). More recently, a study from Egypt investigating protozoal contamination of fruit juices used centrifugation, Sheather's sugar flotation and staining with mZN for analysis, but no indication of recovery efficiency or limits of detection were provided (Mossallam 2010).

One of the first studies to investigate detection of *Cryptosporidium* in milk used PCR methods to detect between 1 and 10 oocysts in 20 mL of artificially contaminated milk (Laberge et al. 1996), while another early method development study also investigated other dairy products, including yoghurt and ice cream (Deng and Cliver 1999). In the latter study, seeding of 100 mL samples of low fat milk with high numbers of *Cryptosporidium* oocysts and analysis by sucrose flotation followed by IFAT for detection resulted in a mean recovery efficiency of over 80 % (Deng and Cliver 1999). Using homogenised milk and pre-labelled oocysts for seeding studies (10 oocysts in 100 mL) with IMS followed by IFAT or PCR for detection resulted in recovery efficiencies of over 95 % and a sensitivity for PCR down to ten oocysts (Deng et al. 2000). Another IMS-PCR seeding trial reported detection of less than ten oocysts (Di Pinto and Tantillo 2002). One study investigated different flotation solutions (sucrose, sodium chloride, magnesium sulphate, zinc sulphate, aluminium sulphate and ammonium sulphate) for isolating *Cryptosporidium* oocysts from milk, with subsequent detection by staining (Kinyoun's technique and Koster's modified technique) and microscopy (Machado et al. 2006). Coagulation of the milk with some of the solutions made the procedure more cumbersome, but recovery efficiencies of over 40 % were reported for magnesium sulphate. Use of PCR for detection of *Cryptosporidium* oocysts in milk was also investigated in a further two studies, one of which compared conventional PCR, real-time PCR and nested PCR and reported detection limits per mL of milk of 10^3, 10^2 and 10 oocysts, respectively (Minarovicová et al. 2007), while the same group reported that microfiltration of milk (using cellulose acetate and cellulose nitrate filters, with pore size, 3.0 µm), followed by elution (sodium pyrophosphate and Tween 80) in a shaker, and then detection by single-tube nested real-time PCR had a detection limit of ten oocysts per 100 mL of milk (Minarovicova et al. 2011).

Despite the proven importance of beverages as vehicles for outbreaks of infection and the various efforts towards method development, with particular emphasis on detection via molecular methods, there are currently no widely known plans to develop a Standard Method for analysis of beverages for contamination with *Cryptosporidium* oocysts.

4.5   Water Used in the Food Industry

The food industry uses a huge volume of water; water is used as an ingredient, as an initial and intermediate cleaning medium, for conditioning raw materials (soaking, cleaning, blanching and chilling) and as a conveyor of raw materials. Prior to food processing, water is used for crop irrigation, for application of chemicals such as pesticides, for depuration and for general washing and cleaning purposes. Although water of non-potable quality may be appropriate for some uses in the water industry, for others uses it is essential that the quality of the water is of the same microbiological quality as drinking water. In order to reduce water usage/wastage, in the fresh produce industry in particular, water might be recycled, often with a purification step (use of a sanitiser) to inactivate pathogens already removed from the produce (Gil et al. 2009). This sanitisation step often involves chlorination, but other technologies such as photocatalysis (Selma et al. 2008a), ozone and UV (Selma et al. 2008b) have also been proposed. One problem with the sanitising step is that it is usually directed towards elimination of bacteria, which may have the potential to multiply on the produce. However, sanitisers that are effective against bacteria may be ineffective against protozoa; although *Cryptosporidium* do not replicate outside the host environment such as in washwater, by reusing such water for a further washing step, oocysts have the potential to be distributed onto previously clean areas or batches of product. This may result in a point contamination becoming spread throughout a whole batch, or among several batches, such that a limited region of contamination that might be associated with the likelihood of a single case of infection may be distributed such that the possibility of a single case expands to the possibility of an outbreak.

Methods specifically directed towards analysing water used in the food industry for *Cryptosporidium* oocysts have not been developed, but approaches based on the standard protocols for drinking water would probably be most appropriate (i.e. US EPA Method 1623; ISO Method 15552) and have indeed been used for both irrigation water and processing water in the food industry (Robertson and Gjerde 2001b; Chaidez et al. 2005; Paruch et al. submitted). It should, however, be realised that reused processing water may have a greater load of contaminating debris than drinking water and thus a lower recovery efficiency may be expected, while irrigation water may be of microbiologically low quality, or even contain untreated wastewater. The use of an internal process control may be of use in such instances to monitor recovery efficiencies (Warnecke et al. 2003).

Chapter 5
Occurrence of *Cryptosporidium* oocysts in Different Food Matrices: Results of Surveys

The results of surveys for pathogens, such as *Cryptosporidium* oocysts, in different food matrices provide a snapshot of what was found on a particular food sample, on a particular occasion, under particular conditions, using a particular method, in a particular laboratory, by a particular analyst. Some surveys do not even provide information regarding the recovery efficiency of the method used, or may only provide information on recovery efficiencies produced in the research group that first developed or published the method. As some of the methods are relatively expensive (particularly IMS, if used) often surveys consist of only a small number of samples, and as recovery efficiencies are often superior with smaller sample sizes, only a small quantity of the product is analysed. Thus, results from such surveys can only be considered to give a very diffuse insight into the risks of ingestion of a *Cryptosporidium* oocyst from a particular product. Furthermore, surveys that do not investigate the species of any *Cryptosporidium* oocysts detected cannot even determine with certainty whether the oocysts that are found are infectious to humans, and thus of public health significance. And surveys that do not consider the viability or infection potential of the oocysts are similarly hampered. Nevertheless, while it is important to acknowledge and be aware of the limitations of such surveys, it should also be realised that these results are our only available verified, scientific handle on contamination of food matrices with potentially infective *Cryptosporidium* oocysts, and thus the information that they provide is useful. In addition, investigation of food matrices for *Cryptosporidium* contamination in an outbreak situation has the potential to identify infection routes, and thus take measures against them. Information obtained from food products analysed during an outbreak event is of greater value if baseline survey data are also already available against which the outbreak-related analyses can be compared. Occurrence data are being improved all

L.J. Robertson, *Cryptosporidium as a Foodborne Pathogen*, SpringerBriefs in Food, Health, and Nutrition, DOI 10.1007/978-1-4614-9378-5_5, © Lucy J. Robertson 2014

the time, as further studies are conducted with better, more efficient methods. The information outlined in the sections below is intended to provide an insight of what we know about the occurrence of *Cryptosporidium* oocysts on different product types as of today—more up-to-date information should always be sought. When comparing occurrence results from different studies, the method used and the recovery efficiency of that method should always be borne in mind, as surveys using very different methods or similar methods but with different recovery efficiencies cannot be readily compared, or such comparisons may give an erroneous impression. When recovery data are not provided, it is probably most appropriate to assume that the efficiency of the method used is low, and thus any contamination data provided are likely to be conservative.

It should be noted that using recovery efficiency data that has been obtained by another research group, even if the analytical methods used are very similar, can be misleading, and extrapolated results may give incorrect insights regarding the extent of contamination. Examination of the literature reveals that different laboratories may achieve very different recovery efficiencies even when using methods that are largely identical; for example, while Cook et al. (2007) reported a recovery efficiency of *Cryptosporidium* cysts from salad leaves of 36.2 ± 19.7 % ($n = 20$) when internal controls were used on samples, a study using the same technique in Spain (Amorós et al. 2010) reported a mean recovery efficiency of *Cryptosporidium* oocysts from salad vegetables (Chinese cabbage and lettuces) of only 24.5 ± 3.5 % ($n = 8$) according to internal controls, and a study from Poland concerned with various vegetables (Rzeżutka et al. 2010) reported recovery efficiencies from internal controls ranging from 4 % (a sample of white cabbage) to 47 % (another sample of white cabbage), with a combined recovery efficiency (10 samples including samples of Brussels sprouts ($n = 1$), cauliflower ($n = 1$), leek ($n = 2$), lettuce ($n = 1$), spring onion ($n = 1$), white cabbage ($n = 3$), Peking cabbage ($n = 1$)) of 23.9 ± 14.0 %.

5.1   Fruits and Vegetables

Outbreaks and method development are probably the two greatest drivers for surveys for contamination of food products with *Cryptosporidium* oocysts, and fresh fruits and vegetables are one of the food matrices for which the greatest number of surveys for *Cryptosporidium* contamination has been conducted. *Cryptosporidium* oocysts have been detected as contaminants on/in a range of raw vegetables and fruit (see Table 5.1) and a widespread, low-level contamination of fresh produce can be generally inferred from these results. However, as previously noted, differences in recovery efficiencies obtained by different analytical procedures or even by similar analytical procedures used by different laboratories mean that comparison of results obtained in different studies is problematic. Of the studies listed in Table 5.1, only those from Europe use analytical methodologies upon which the principles adopted by the proposed ISO standard (ISO Draft 18744) are based. Although survey data are available from Europe, North America (Canada), Asia, and Central and South America, surveys from Africa are largely lacking (one survey from Egypt). This is unfortunate as although zoonotic cryptosporidiosis may be less common in African countries, the impact of cryptosporidiosis on different African populations is well recognised (e.g. Kotloff et al. 2013), and obtaining information on probable transmission routes would be advantageous. It is also surprising that surveys for *Cryptosporidium* contamination in fresh produce, such as salad vegetables and fresh fruit, in North America have only been conducted in Canada (none from the USA), and there are also a lack of such surveys from Australasia.

Some of the studies listed in Table 5.1 have used molecular methods for detection (sometimes in combination with other methods such as IFAT). One disadvantage of using molecular methods is that quantification of oocysts is less straight-forward (in one study, Dixon et al. (2013), in which quantification was possible, although not undertaken, as IFAT was used on PCR-positive samples, the authors state that in "the majority of surveillance studies" no attempt is made to enumerate the parasites; however, this is manifestly not the case for those studies using methods based on the principles adopted by the ISO Method; see Table 5.1). Information on the concentration of oocysts detected per serving or quantity of a specific product is useful input data for risk assessment, and can be obtained with very little extra effort when IFAT is used as the detection method.

However, molecular methods do provide useful information on species or subtype of oocysts that are detected; this information may not only indicate the possible public health importance of the oocysts detected but also provide clues about the likely source of contamination. In one of the Canadian studies (Dixon et al. 2013), of the 29 samples for which sequences were obtained, all were found to be of *C. parvum* (subtyping information, obtained, for example, by molecular analysis at the gp60 gene, was not sought, although the authors mention that further molecular subtyping would be needed for further source tracking). In the study from Poland (Rzeżutka et al. 2010) although IFAT was used for detection, downstream molecular processing was used on oocysts retrieved from the microscope slides, and the oocysts from the celery sample were identified as *C. parvum*. Although subtyping at the gp60 locus was attempted it was not successful. One point made in this study is that the use of an internal process control to determine recovery efficiency abrogates the possibility of downstream genotyping. Thus the investigators must be sure of the intention of their survey and use the methods that are most appropriate; it should be noted that methods that are best suited for obtaining background survey information may be less suitable for investigating contamination of produce in an outbreak situation. Other studies that have used PCR as an identification method (e.g. Rai et al. 2008; Bohaychuk et al. 2009) have not attempted to obtain other species or subtype data.

Whether the oocysts detected are infectious/viable has not been explored in any study to date, presumably because the oocysts tend to be found at low concentrations (this is the reason stated for not exploring viability in the study by Dixon et al. 2013) and also, for many studies, the detection method inactivates them or reduces their viability (e.g. fixing to microscope slides).

Table 5.1 Occurrence of *Cryptosporidium* oocysts on fresh produce[a]

Location of survey and origin of produce analysed	Type of fresh produce analysed	Results: proportion of samples contaminated with *Cryptosporidium* oocysts	Concentrations of oocysts on positive samples	Reference
Peru; locally produce obtained from small markets	110 samples comprising: basil, cabbage, celery, cilantro, green onions, green chili, herbs, leeks, lettuce, parsley	Overall prevalence of 14.5 %	No data provided	Ortega et al. (1997)
Norway; produce both imported and locally produced (beansprouts grown locally from imported seed)	475 samples comprising: alfalfa sprouts, dill, lettuce, mung beansprouts, mushrooms, parsley, precut salad mix, radish sprouts, raspberries, strawberries	Lettuce: 4 % mung beansprouts: 9 % *Cryptosporidium* oocysts not detected in other samples	No. oocysts per 100 g: lettuce: 1–6; mung beansprouts: 2–6	Robertson and Gjerde (2001b)
Costa Rica; produce obtained from agricultural markets	250 samples comprising: lettuce, parsley, cilantro, strawberries, blackberries	Dry (wet) season results: Lettuce 24 (4) %; parsley 4 (0) %; cilantro 4 (0) %; blackberries 8 (4) % All strawberry samples negative	No data provided	Calvo et al. (2004)
Kathmandu, Nepal; produce obtained from 4 markets	30 samples of each of: leaves of radishes, cauliflower, green onions, cabbages, mustard leaves, carrots	Radish: 17 % Cauliflower: 0 Cabbage: 13 % Mustard leaves: 3 % Other samples negative	No data provided	Ghimire et al. (2005)
Palermo, Sicily; produce from local retailers	20 raw vegetable mixes (leafy vegetables and carrots), 20 leafy vegetable mixes	*Cryptosporidium* oocysts not detected	–	Di Benedetto et al. (2007)
York, UK; produce from local retailers	20 raw vegetable mixes including: various lettuce varieties, carrots, mange tout, spring onions, parsley, chilies, baby sweet corn, asparagus	*Cryptosporidium* oocysts not detected	–	Cook et al. (2007)
Cambodia; harvested from 3 locations in same lake	36 samples of water spinach	17 % prevalence	Average oocyst concentration per gram: 0.5	Vuong et al. (2007)
Northern India; samples from produce transported to market	9 samples comprising: amaranth leaves, amaranth roots, carrot roots, mint leaves, spinach leaves, cabbage, lettuce, radish, chili fruits	Only amaranth leaf sample positive for *Cryptosporidium*	No data provided	Rai et al. (2008)
Canada; samples from farmers market	157 samples comprising: lettuce, spinach, green onions, strawberries	1 spinach sample positive	No data provided	Bohaychuk et al. (2009)

Location/source	Samples	Results	No. oocysts	Reference
Valencia, Spain; produce from agricultural settings	19 samples comprising Chinese cabbage, lollo rosso lettuce, romaine lettuce	Chinese cabbage: 33 %; Lollo rosso lettuce: 75 %; Romaine lettuce: 78 %	No. oocysts per 50 g: Chinese cabbage: 4–7; lollo rosso lettuce: 6–15; Romaine lettuce: 2–10	Amorós et al. (2010)
Poland; produce from farmers markets	163 samples comprising: Peking cabbages, leeks, white cabbages, red cabbages, lettuces, spring onions, celery, cauliflowers, broccoli, spinach, Brussels sprouts, raspberries, strawberries	Overall prevalence of 3.6 %, with oocysts detected in 1 leek sample, 1 celery sample, 2 cabbage samples, 1 red cabbage sample, and 1 Peking cabbage sample	No. oocysts per 30 g: leek: 1; celery: 47; cabbage: 1; red cabbage: 2; Peking cabbage: 4	Rzeżutka et al. (2010)
Pathumthani Province, Thailand; samples collected in the field	3 samples (lettuce, washed lettuce, and water spinach)	All samples positive by at least one method	No. oocysts per 200 g: lettuce: 1.2; washed lettuce: 1.3; water spinach: 4–10 (depending on analytical method)	Keserue et al. (2012)
Alexandria, Egypt; samples collected from various vendors	300 samples comprising: lettuce, rocket, parsley, leek, and green onion	Lettuce: 43 %; Rocket: 45 %; Parsley: 33 %; Leek: 13 %; Green onion: 12 %	No data provided	El Said Said (2012)
Ontario, Canada; both imported and locally produced	544 samples comprising: iceberg lettuce, romaine lettuce, baby lettuces, leaf lettuce, radicchio, endive, escarole, spinach	Overall prevalence of 5.9 %	No data provided	Dixon et al. (2013)
Norway; both imported and locally produced	41 samples comprising: salad leaves (of different varieties), raspberries, and mange tout	Cryptosporidium oocysts not detected	–	Johannessen et al. (2013)
Tehran, Iran; samples collected from vegetable farms from 5 different regions close to Tehran	496 samples comprising: mint, leeks, cress, green onion, coriander, basil	Mint: 8.5 %; Leek: 3.3 %; Cress: 8.9 %; Green onion: 14.8 %; Coriander: 6.7 %; Basil: 1.1 %	No data provided	Ranjbar-Bahadori et al. (2013)

[a]Adapted and updated from information previously published in Robertson and Chalmers (2013) and Robertson and Fayer (2012)

5.2   Shellfish

Experimental studies have clearly demonstrated that shellfish (bivalve molluscs) have the ability to filter out, retain, and accumulate *Cryptosporidium* oocysts from the surrounding water (Graczyk et al. 2003). Furthermore, *Cryptosporidium* oocysts have been shown to associate with marine macroaggregates (Shapiro et al. 2013). Derivations of enrichment factor estimations from lab-scale experiments have demonstrated that oocysts are two to three orders of magnitude more concentrated in aggregates than in estuarine and marine waters surrounding the aggregates, and this has been speculated to enhance their bioavailability to invertebrates and thus their subsequent incorporation into the marine food web (Shapiro et al. 2013).

Although no outbreaks of cryptosporidiosis associated with shellfish have been reported, several surveys of shellfish for *Cryptosporidium* oocysts have been published. Some of these surveys have been conducted with the intention of investigating oocyst accumulation in shellfish as a method of assessing contamination in water (e.g. Izumi et al. 2006; Lucy et al. 2008) or even as a means of depleting oocysts from sewage to prevent further contamination of the aquatic environment (Izumi et al. 2012). However, most of the surveys have been conducted with consideration as shellfish as a potential vehicle for foodborne transmission (see Table 5.2), and, intriguingly, there have been more surveys of shellfish than of vegetables or fresh produce, despite the latter having been associated with proven outbreaks. While there have been no surveys of fresh produce for *Cryptosporidium* contamination documented from the USA to date, for shellfish five surveys have been documented from the USA, as well as several surveys from Europe and surveys from South America, North Africa, and Asia.

As with surveys of fresh produce, the variation in analytical methods means that comparison between surveys is difficult. Also, ascertaining the proportion of positive samples can be problematic for some studies as samples may be prepared in different ways (single molluscs, groups of molluscs, different sites), and similarly for concentrations of oocysts. This lack of similarity between surveys is reflected in the data presented from 32 studies in Table 5.2, in which only 7 of the 28 (25 %) studies with positive results attempt to quantify the number of oocysts detected. In comparison, of the 13 studies concerned with fresh produce listed in Table 5.1, of the 10 that had positive results, 5 (50 %) attempted to quantify the extent of contamination (oocysts per gram of produce). The lack of quantification in shellfish surveys may also be partly due to several of the shellfish studies using molecular methods for detection, although various studies recommend that molecular methods and microscopy-based methods are used to complement each other (e.g. Fayer et al. 2003; Giangaspero et al. 2005).

Furthermore, and also in contrast with the surveys conducted on fresh produce, several of the shellfish surveys have attempted to identify the viability/infectivity of oocysts detected in shellfish, using a range of techniques. For example, mouse infectivity studies were used by Fayer et al. (1998), Gómez-Bautista et al. (2000), and Fayer et al. (2002) and, in these studies it was demonstrated that the oocysts

Table 5.2 Occurrence of *Cryptosporidium* oocysts in shellfish potentially destined for human consumption

Location of survey	Shellfish analysed	Results: proportion or number of samples contaminated with *Cryptosporidium* oocysts	Oocyst concentrations in positive samples	Reference
Three sites, Sligo, Republic of Ireland	Common mussel (*Mytilus edulis*)	3/26 (11.5 %)	No data provided	Chalmers et al. (1997)
Maryland tributaries of Chesapeake Bay, USA	Eastern oyster (*Crassostrea virginica*)	142/360 (39.4 %)	No data provided	Fayer et al. (1998)
9 sites on the Gallacian coast, Northwest Spain	Mediterranean mussel (*Mytilus galloprovincialis*); common cockle (*Cerastoderma edule*)	31 mussels and 6 cockles positive	No data provided	Gómez-Bautista et al. (2000)
Galician coast, Northwest Spain and imported into Spain from Italy (*R. philippinarum*) and UK (1 sample *O. edulis*)	Mediterranean mussel (*M. galloprovincialis*); Manila clam (*Ruditapes philippinarum*); Pullet carpet shell (*Venerupis pullastra*); Smooth Artemis (*Dosinia exoleta*); European flat oyster (*Ostrea edulis*); banded carpet shell clam (*Venerupis rhomboideus*); warty venus shell (*Venus verrucosa*)	Mussels (pooled homogenate): 6/15 (40.0 %). Clams (pooled homogenate): 10/17 (58.8 %). Oysters (pooled homogenate): 5/6 (83.3 %)	No data provided	Freire-Santos et al. (2000)
Belfast Lough, N. Ireland, UK	Common mussel (*M. edulis*)	2/16 (12.5 %)	No data provided	Lowery et al. (2001)
Chesapeake Bay, USA	Eastern oyster (*C. virginica*)	311/1,590 (19.6 %)	No data provided	Fayer et al. (2002)
37 commercial harvesting sites on Atlantic Coast, USA	Eastern oyster (*C. virginica*); Clams (species not stated)	Oysters: 32/550 (5.8 %) and 21/110 (19.1 %) depending on method. Clams: 3/375 (0.8 %) and 12/75 (16.0 %) depending on method	No data provided	Fayer et al. (2003)
Commercially available in markets in Alexandria, Egypt	Gandoffli (*Caelatura pruneri*), wedgeshell clam (Om el Kholool) (*Donax trunculus limiacus*)	Detected in both species in 2 out of 3 batches	No data provided	Negm (2003)
9 sites on the Galician coast, Northwest Spain and shellfish imported into Spain from Italy, UK, Ireland, and Portugal	Mediterranean mussel (*M. galloprovincialis*); common cockle (*C. edule*); carpet shell clam (*Ruditapes decussatus*); Manila clam (*R. philippinarum*); Pullet carpet shell (*V. pullastra*); Smooth Artemis (*D. exoleta*); European flat oyster (*O. edulis*)	Mussels (pooled homogenate): 35/107 (32.7 %); cockles (pooled homogenate): 5/24 (20.8 %); clams (pooled homogenate): 20/68 (29.4 %); oysters (pooled homogenate): 23/42 (29.4 %)	No data provided	Gómez-Couso et al. (2003a)

(continued)

Table 5.2 (continued)

Location of survey	Shellfish analysed	Results: proportion or number of samples contaminated with *Cryptosporidium* oocysts	Oocyst concentrations in positive samples	Reference
Specimens originated from Spain, UK, Italy, Ireland, New Zealand	Mediterranean mussel (*M. galloprovincialis*); common cockle (*C. edule*); Manila clam (*R. philippinarum*); Pullet carpet shell (*V. pullastra*); Smooth Artemis (*D. exoleta*); European flat oyster (*O. edulis*); banded carpet shell clam (*V. rhomboideus*); warty venus shell (*V. verrucosa*); green lipped mussel (*Perna canaliculus*)	Spanish samples: mussels (pooled homogenate): 12/22 (54.6 %) and 8/22 (36.4 %) depending on method; clams (pooled homogenate): 10/18 (55.6 %) and 9/18 (50 %) depending on method; oysters (pooled homogenate): 6/9 (66.7 %) and 5/9 (55.6 %) depending on method UK samples: mussels (homogenate): 3/20 (15 %); cockles (homogenate): 1/18 (5.6 %)	No data provided	Gómez-Couso et al. (2004)
Coast of Adriatic Sea, Italy (mouths of 4 rivers in Abruzzo region)	Striped venus clam (*Chamelea gallina*)	2/4 pools (sites) positive (50 %)	No data provided	Traversa et al. (2004)
Nine sites, California coast, USA	California mussel (*Mytilus californianus*); Mediterranean mussel (*M. galloprovincialis*)	19/156 (12.2 %)	No data provided	Miller et al. (2005)
Coast of Adriatic Sea, Italy (mouths of 4 rivers in Abruzzo region)	Striped venus clam (*C. gallina*)	23/32 (71.9 %)	8–45 oocysts per gram in tissue homogenates and from 18 to 200 oocysts/mL in the haemolymph	Giangaspero et al. (2005)
Galician coast, Northwest Spain	Mediterranean mussel (*M. galloprovincialis*)	Pooled homogenate: 54/222 (24.3 %) and 28/222 (12.6 %) depending on method. Combined total: 69/222 (31.1 %)	No data provided	Gómez-Couso et al. (2006a)
Four estuaries, Galician coast, Northwest Spain	Mediterranean mussel (*M. galloprovincialis*)	Pooled homogenate: 42/184 (22.8 %) and 26/184 (14.1 %) depending on method. Combined total: 54/184 (29.3 %)	No data provided	Gómez-Couso et al. (2006b)
Three sites, Normandy coast, France	Common mussel (*M. edulis*)	11/11 (100 %)	0.05–0.90 oocysts/mussel	Li et al. (2006)
St Lawrence River, Québec, Canada	Softshell clam (*Mya arenaria*)	29/41 (70.7 %)	Data not provided	Lévesque et al. (2006)
Guadiana river hydrographic basin, southeast Portugal	Freshwater clams: *Anodonta anatina*, *Unio pictorum*, *Corbicula fluminea*	*U. pictorum*: 7/24 (29 %); *A. aratina*: 4/12 (33 %); *C. fluminea*: 7/19 (37 %)	1–8 oocysts/25 µL of resuspended pellet (approximately equivalent to 0.5–2.5 shellfish)	Melo et al. (2006)
Oosterschelde, Netherlands	Pacific cupped oyster (*Crassostrea gigas*)	11/179 (6.1 %)	Data not provided	Schets et al. (2007)

Location	Species	Prevalence	Concentration / load	Reference
53 commercial harvesting sites in Chesapeake Bay, USA	Eastern oyster (*C. virginica*)	235/265 (88.7 %)	1–125 oocysts per 6 oysters, depending on site. Average oocyst load per 6 oysters ranged from 22 to 80, depending on site, with over cumulative mean of 42 oocysts per 6 oysters	Graczyk et al. (2007)
3 clam farms, Adriatic coast, Italy	Manila clam (*R. philippinarum*)	Varied according to site (infection rates of 0 %, 0.14 %, and 1.15 %)	Data not provided	Molini et al. (2007)
13 different sites on the Norwegian coast	Blue mussel (*M. edulis*); horse mussel (*Modiolus modiolus*); European flat oyster (*O. edulis*)	Blue mussels: 6/14 (43 %); horse mussels: 0/1 (0 %); oysters: 0/1 (0 %)	Blue mussels: 1–2 oocysts per sample (approximately 3 g)	Robertson and Gjerde (2008)
Coastal area from São Paulo, Brazil	Mangrove cupped oyster (*Crassostrea rhizophorae*); cockle (*Tivela mactroides*)	Oysters: 10.0 % of gill pools; cockles: 50.0 % of gill pools	Mean concentration estimates: 12 oocysts per oyster and 60 oocysts per cockle	Leal Diego et al. (2008)
Sligo Bay, Ireland	Blue mussel (*M. edulis*)	Sample found positive	2.3 oocysts per gram	Lucy et al. (2008)
Mali Ston Bay, Adriatic Sea, Croatia	Mediterranean mussel (*M. galloprovincialis*)	31/184 samples (16.8 %)	Data not provided	Mladineo et al. (2009)
Shellfish markets, Bangkok and Samut, Thailand	Green mussel (*Perna viridis*)	Bangkok: 4/24 (8.3 %); Samut: 5/32 (15.6 %)	Data not provided	Srisuphanunt et al. (2009)
Varano Lagoon, Southern Italy	Mediterranean mussel (*M. galloprovincialis*); grooved carpet shell clam (*R. decussatus*)	Mussels: 0/24 pools (0 %); Clams: 0/23 pools (0 %)	–	Giangaspero et al. (2009)
Nunavik, Quebec	Blue mussel (*M. edulis*)	8/11 samples (73 %)	Data not provided	Lévesque et al. (2010)
Varano Lagoon, Southern Italy	Mediterranean mussel (*M. galloprovincialis*); grooved carpet shell clam (*R. decussatus*); Pacific cupped oyster (*C. gigas*)	0/38 pools (0 %)	–	Francavilla et al. (2012)
4 sites Florianópolis, Santa Catarina State, Brazil	Pacific cupped oyster (*Crassostrea gigas*)	1/4 sites (25 %)	Data not provided	Souza et al. (2012)
"Vale do Ribeira", Cananéia city, Brazil	Oysters (*Crassostrea brasiliana*)—both at collection and following depuration	At collection: 0/11 (0 %); After depuration: 0/11 (0 %)	–	Leal Diego et al. (2013)
Shellfish farms on western and north-eastern coasts of Sardinia	Mediterranean mussel (*M. galloprovincialis*); Pacific cupped oyster (*C. gigas*)	Mussels: 0/72 pools (0 %); Oysters: 0/60 pools (0 %)	–	Tedde et al. (2013)

recovered from shellfish were infectious; a further study by Fayer et al. (2003) also used a bioassay for investigating infectivity of *Cryptosporidium* oocysts detected in shellfish, but did not result in infections establishing in mice. Gómez-Couso et al. (2003a) used vital dye inclusion/exclusion as an indicator of viability and showed that around 50 % of samples contained viable oocysts, while Graczyk et al. (2007) applied a fluorescent in situ hybridisation (FISH) assay as a measure of viability and determined that 83 % of the oyster batches contained viable oocysts, with a mean of 28 viable oocysts per contaminated group of 6 oysters. However, seeding studies into shellfish have demonstrated that FISH can overestimate viability as compared with in vitro excystation (Robertson and Gjerde 2008), presumably due to the stability of rDNA within oocysts, as has previously been described.

While only two of the surveys of fresh produce for *Cryptosporidium* contamination used molecular analysis to obtain information on the species detected, several of the surveys for shellfish have attempted to obtain information on contaminant species, although not all these investigations have been successful. Nevertheless, the most common species of *Cryptosporidium* identified as a contaminant in shellfish seems to be *C. parvum* (Gómez-Bautista et al. 2000; Fayer et al. 2002, 2003; Traversa et al. 2004; Gómez-Couso et al. 2004, 2006a, c; Miller et al. 2005; Giangaspero et al. 2005; Leoni et al. 2007; Molini et al. 2007). In addition, several studies have reported contamination of shellfish with *C. hominis* (Lowery et al. 2001; Fayer et al. 2002, 2003; Gómez-Couso et al. 2004; Miller et al. 2005; Molini et al. 2007). A few studies have also reported the occasional occurrence of *Cryptosporidium* that may have less association with human infection, including *C. baileyi* (Fayer et al. 2002), *C. meleagridis* (Fayer et al. 2003), *C. andersoni* (Gómez-Couso et al. 2004), and *C. felis* (Miller et al. 2005).

5.3 Meat

Currently there are few reports on investigation of meat for *Cryptosporidium* oocyst contamination. A study from Ireland investigated the contamination of beef carcasses for *Cryptosporidium* contamination; of 288 carcass meat samples analysed during the course of a year, none were found to be contaminated with *Cryptosporidium* oocysts although oocysts were present in the faeces of some of the cattle presented for slaughter during the study (Duffy et al. 2003). A study from India investigated three batches of goat meat for contamination; although none were considered positive by IFAT, one was reported to be positive by PCR (Rai et al. 2008). Although PCR was used for successful detection, the authors do not explore the species of *Cryptosporidium*, and therefore it is difficult to speculate on whether contamination of goat meat is likely to be from faecal matter from the goats themselves, or from human handlers of the meat. Presumably contamination of meat via flies or other vectors is also possible.

In another unpublished study, contamination with *Cryptosporidium* was detected in various raw meats (chicken breasts, minced beef, pork chops) from retail outlets using PCR for detection, and one minced beef sample was also found to be positive for *Cryptosporidium* by IFAT (reported in Dixon 2009). Sequencing of all the PCR products indicated that the *C. parvum* was the contaminating species for all products (Dixon 2009); although *C. parvum* is zoonotic, the fact that the contamination with *C. parvum* was not only in beef but also in chicken and pork (where other species of *Cryptosporidium* are more common) suggests that at least some of the contamination could have been due to human handling. It is probably unlikely that chickens are slaughtered in the same facilities as cattle, and thus between-animal product contamination at the slaughterhouse also seems unlikely.

Finally, a study from Scandinavia investigated possible contamination of cured hams subsequent to an extensive outbreak of waterborne cryptosporidiosis (Robertson and Huang 2012), in which contact between the contaminated water and the meat products could have occurred. A total of 11 samples were analysed from two different products; in one sample an object resembling a *Cryptosporidium* oocyst was identified, but was recorded as putative due to the absence of internal contents and some deformity.

5.4   Beverages

Given that both milk and fruit juices have been associated with documented outbreaks of cryptosporidiosis, it might be expected that beverages have been surveyed relatively widely for contamination with *Cryptosporidium*. However, only two surveys investigating contamination of milk with *Cryptosporidium* were identified, and only two prospective surveys investigating contamination of fruit juices, although juice samples have been analysed in association with juice-related outbreaks. Interestingly, all the studies on contamination of beverages with *Cryptosporidium* in the absence of a foodborne outbreak have been conducted in less-developed countries (Trinidad, India, Egypt, Ethiopia). This is probably because the majority of milk, fruit juices, and other beverages available in industrialised countries have been pasteurised and therefore are usually not considered to pose a risk of infection to consumers.

One milk survey was conducted on bulk milk samples from dairy farms in Trinidad; this survey, which was based upon formol-ether sedimentation and acid fast staining for detection (without presenting any data regarding the efficiency of the method), failed to detect any *Cryptosporidium* contamination in 177 samples (Adesiyun et al. 1996). The second study was from India and did not detect any *Cryptosporidium* contamination in three batches of milk (Rai et al. 2008). The method used in this survey included sedimentation, acid fast or Lugol's iodine microscopy and immunofluorescence microscopy, and SSU rRNA gene PCR, and again no efficiency data regarding the method are presented.

A study from Egypt investigated a range of different juices (strawberry, sugar cane, mango, lemon, and orange) for contamination with protozoan parasites, and reports the detection of *Cryptosporidium* oocysts in all juice types (Mossallam

2010). For each fruit juice, samples were collected from roadside stalls for analysis, and the occurrence of *Cryptosporidium* contamination ranged from 29 % (10 samples positive) for strawberry juice and lemon juice to 14 % (5 samples positive) for sugar cane juice. In total, 38 samples of juice (22 %) were reported to contain *Cryptosporidium* oocysts, but oocyst numbers per volume of sample analysed are not provided; such information would be relevant and useful for risk assessment and it is unclear why this information is not included in the report. This study also attempted to assess oocyst viability and infectivity, both by examining inclusion and exclusion of the fluorogenic dyes fluorescein diacetate and propidium iodide, and by mouse infectivity trials. The results from these studies indicated low oocyst viability and lack of infectivity in the most acidic juices (orange (pH 2.9) and lemon (pH 3.2)), but high viability and infectivity in the other three juices that were less acid (ranging from pH 4 (mango) to pH 7.5 (sugar cane)). The reduced survival of *Cryptosporidium* oocysts in beverages with a low pH has previously been reported (Friedman et al. 1997).

In an experimental study on the capacity of filth flies to contaminate mango juice kept in the vicinity of calves diagnosed with *Cryptosporidium* infection, after 2 h in which dishes of mango juice were exposed to access by flies in the environment, 2 out of 10 samples of mango juice were contaminated, and, after 8 h, 6 dishes had been contaminated with oocysts (Fetene et al. 2011).

5.5　Water Used in the Food Industry

5.5.1　*Irrigation Water, etc.*

Irrigation with untreated water is a potential route of crop contamination with *Cryptosporidium* oocysts. Surface waters may contain *Cryptosporidium* oocysts, either from sewage discharge or from contamination from animal sources, and if this water is used for irrigation or for other agricultural uses (e.g. application of pesticides or fertilisers), then these may be transferred onto the surfaces of the crops. In addition, in developing countries, or countries where water resources are scarce, it makes sense to use wastewater for crop irrigation. It has been estimated that around 20 million ha globally are irrigated by raw, treated, and/or partially diluted wastewater (Hamilton et al. 2006a), and it has been shown that the use of untreated wastewater in agriculture can have major financial and nutritional benefits for farmers and consumers (Ensink and van der Hoek 2009), although the risk of intestinal disease was elevated (although cryptosporidiosis infection was apparently not specifically elevated). Thus, the potential for transfer of pathogens, including *Cryptosporidium* oocysts, from irrigation water, especially when it includes wastewater, to crops should not be overlooked. In a recent study from Iran, crops that were irrigated with wastewater had a statistically significantly greater likelihood of being contaminated with *Cryptosporidium* oocysts, than crops that were irrigated with well water (Ranjbar-Bahadori et al. 2013). However, a wastewater pretreatment step may make a large difference to the contamination potential. For example, a study in which wastewater was used for irrigation of vegetable crops identified *Cryptosporidium* contamination of both lettuce and fennel samples (Lonigro et al. 2006), but when the wastewater was filtered effectively, the irrigation water was not found to provide a contamination risk. A study from Thailand has also suggested that flow-through canals, which can be viewed as waste stabilisation ponds, remove *Cryptosporidium* oocysts at the generalised specific rate of 0.3, with higher removal rates during the rainy season, and with the main removal mechanism considered to be predation (Diallo et al. 2009). Nevertheless, although the concentration of pathogens in sewage decreases successively at each treatment step, if the initial concentrations are high, then the final effluents may still contain a large number of these pathogens, including *Cryptosporidium*, and these may pose a serious public health risk (Armon et al. 2002).

It should also be remembered that using untreated wastewater in agriculture may carry a risk to the farmers themselves as they are going to be in closer contact with a medium that is likely to contain pathogens. One study from Eritrea suggested that agricultural use of untreated wastewater was the major cause of the increase in gastrointestinal diseases (Srikanth and Naik 2004). Note that unhygienic post-harvest handling (such as washing in contaminated water) might increase, rather than decrease, contamination levels (Ensink et al. 2007). In an outbreak of cryptosporidiosis associated with apple cider, well water used for washing the apples prior to pressing the apples was found to be contaminated and may have been the original source of contamination (Anonymous 1997).

The method of irrigation with potentially contaminated water may facilitate or impede potential contamination of produce. For example, sprinkler irrigation, in which the water is piped to one or more central locations within the field and distributed by overhead high-pressure sprinklers or guns, is likely to result in contamination of leaf-crops or fruit produce, whereas drip irrigation or trickle irrigation, in which water is delivered, drop by drop, at or near the root zone of plants, is less likely to result in contamination of leaves and fruits. Additionally, this method is probably more water-efficient, as evaporation and runoff are minimised. Sub-irrigation, in which the water table is artificially raised so that the soil is moistened from below the plants' root zone, is also unlikely to result in contamination of fruit or leaves above ground. Thus, ensuring that irrigation water is delivered to the roots, rather than coming into contact with the leaves and fruit of the plant, is likely to minimise contamination. However, a field-study comparison of two irrigation methods: surface and subsurface of field crops and follow-up of *Cryptosporidium* oocysts in soil at different depths did not demonstrate a clear pattern of distribution (Armon et al. 2002). Studies comparing the effectiveness of simple irrigation methods in reducing microbial contamination of lettuce irrigated with polluted water in urban farming in Ghana have demonstrated that irrigation with drip kits resulted in the lowest levels of contamination (Keraita et al. 2007); however, *Cryptosporidium* was not included in these studies and it was noted also that drip kits often became clogged, required lower crop densities, and restricted other routine farm activities. While the principle of avoiding fruit and leaf contamination may be appropriate both for irrigation and application of fertilisers, it is probably unsuitable for application of pesticides, and for root vegetables, soil contamination itself is also likely to be important. Additionally, for crops growing close to the ground (such as lettuce or strawberries), the potential for splash-up from contaminated soil may also be important. It should be noted that the majority of stomata are on the abaxial side of leaves, and as this is the side that is most likely to become contaminated during splash-up this can be of relevance; studies have demonstrated that *Cryptosporidium* oocysts may become internalised through such pores on spinach leaves (Macarisin et al. 2010a), and this protects the oocysts from environmental degradation and also from removal by washing.

Contamination of crops with *Cryptosporidium* oocysts may occur during irrigation, but should also be considered during intense rain. The potential for contamination of crops during, for example, flooding of fields during extreme weather events is one aspect of climate change that is beginning to be explored in research projects and also as a result of specific events (e.g. Casteel et al. 2006). However, the effect of extreme weather events on contamination of the drinking water supply with pathogens, including *Cryptosporidium*, is presently of greater focus (Cann et al. 2013) than contamination of produce during or following extreme weather events; data from Germany indicate that *Cryptosporidium* levels tend to rise, often significantly, during specific weather events (Kistemann et al. 2002). A small-scale study of irrigation water in Norway demonstrated a higher level of contamination with *Cryptosporidium* oocysts after rainfall (Paruch et al. submitted).

In another study from Norway, samples from a river used for irrigation of lettuces were found to contain *Cryptosporidium* oocysts, with 5 out of 11 samples analysed positive (Robertson and Gjerde 2001b). However, the concentrations of oocysts were low (maximum of 2 oocysts per 10 L) and oocysts were not detected on lettuces from this location that were analysed for contamination. However, a survey from Spain (Amorós et al. 2010) that investigated both water samples from an irrigation canal and crops being irrigated with that water found high concentrations of *Cryptosporidium* oocysts (around 50 oocysts per litre) and also contamination of the vegetable crops irrigated with that water. These high concentrations of oocysts in irrigation water probably reflect inputs into the water; in this case the irrigation canal was known to receive wastewater. Another study investigating irrigation water in Spain (Gracenea et al. 2011) also reported a high prevalence of *Cryptosporidium*, with over 18 % of the 54 samples analysed containing *Cryptosporidium* oocysts, and with concentrations ranging between 3 and 82 oocysts per litre; both runoff from fields with animals and wastewater discharges into rivers feeding the irrigation channels were considered as potential sources of contamination.

A survey of irrigation waters at 3 sites, where fruit and vegetable crops were produced in the USA, and at 22 sites in three Central American countries demonstrated an overall prevalence of *Cryptosporidium* contamination of 36 % (Thurston-Enriquez et al. 2002). While the concentrations of *Cryptosporidium* oocysts detected in irrigation waters in the USA were approximately at the same levels as that reported from Norway low (mean of 25 cysts per 100 L), in Central America the concentrations of oocysts detected were considerably higher, with a mean of over

220 oocysts per 100 L, indicating not only considerable contamination of the irrigation water but, more importantly, a considerable potential for contamination of crops (Thurston-Enriquez et al. 2002). In Mexico also the levels of *Cryptosporidium* oocyst contamination of irrigation waters have been shown to be high, with 48 % of the samples of irrigation water analysed found to contain *Cryptosporidium* oocysts, and with concentrations ranging from under 20 oocysts per 100 L to 200 oocysts per 100 L (Chaidez et al. 2005). Another study from a major irrigation system in Mexico, in which 6 irrigation water samples were analysed for *Cryptosporidium* oocysts, the mean concentration was 1.78 oocysts per 100 mL (maximum concentration detected=5.8 oocysts per 100 mL) despite there being no direct sewage discharges into the irrigation canal system, indicating that contamination was probably from animals, weather events, or localised contamination events (Gortáres-Moroyoqui et al. 2011). A study from Tunisia investigated the occurrence of *Cryptosporidium* in treated wastewater from 18 treatment plants, of which the discharge from 12 of these was destined for irrigation purposes (Ben Ayed et al. 2012); *Cryptosporidium* was detected in effluent from all treatment plants, but it was not clear if the effluent would be diluted before irrigation use, and no attempt was made to sample crops that had been irrigated with this effluent. Water spinach grown near to wastewater discharge outlets near Phnom Penh, Cambodia, has been found to have relatively high contamination with *Cryptosporidium* oocysts, being detected on 17 % of samples at concentrations of 0.5 oocysts per gram (Vuong et al. 2007); however, no effort was made to quantify the *Cryptosporidium* content of the irrigation water or wastewater. The use of sewage-contaminated water for irrigation is common practice in Egypt, and one of the main water sources used for irrigation in Alexandria is the El Mahmoudeya canal, in which a high rate of contamination with *Cryptosporidium* oocysts has been detected (El Said Said 2012). However, no direct attempt to associate contamination of this irrigation source and contamination of fresh produce irrigated from this source has been published to date.

A publication from a study in Thailand (Keserue et al. 2012) suggests that the researchers were able to "track" contaminating oocysts from wastewater to irrigation waters and finally to confirm the contamination of salads and water vegetables. In this study, concentrations of *Cryptosporidium* in irrigation waters were around 10 oocysts per litre, and between 1 and 10 oocysts per 200 g of fresh produce (lettuce or water spinach). The data suggest that a brief wash of the lettuce by farmers prior to sale had only a very minor impact on removing the contamination.

Also in the fresh produce industry, low-level *Cryptosporidium* oocyst contamination was detected in water used in beansprout production in Norway (Robertson and Gjerde 2001b) (contamination assumed to have originated from the beansprout seeds), while in Mexico 16 % of wash-water tank samples used in a packinghouse were found to be contaminated with *Cryptosporidium* oocysts, with concentrations exceeding 100 oocysts per 100 L for some of the samples (Chaidez et al. 2005). Thus, spread of *Cryptosporidium* contamination between produce in produce washing facilities, particularly when wash water is recycled between batches, is one potential aspect for contamination that should be considered as part of Hazard Analysis and Critical Control Point (HACCP) routines (see Chap. 7).

5.5.2   Depuration Water

Depuration processes are another means by which water may be used to spread contamination from contaminated food, here shellfish, to non-contaminated food. During depuration, harvested shellfish are placed in a controlled aquatic environment, where the intention is that any pathogens will be removed by the shellfish purging themselves of their gastrointestinal contents, and thereby any pathogens. The literature investigating whether depuration may result in spread of contamination with *Cryptosporidium* cysts, rather than reduction, is not extensive and some studies provide contrasting results; nevertheless, the accumulated data suggest that the depuration times for *Cryptosporidium* vary between shellfish, and may perhaps result in the spread of contamination among shellfish at depuration plants (Robertson 2007; Willis et al. 2013). For example, Gómez-Bautista et al. (2000) suggested that if a depuration process of 72 h were adopted, then mussels would be able to purge themselves successfully of *Cryptosporidium* oocysts, while, in contrast, experiments on depuration in a number of shellfish species (including mussels, oysters, clams, and cockles) suggested that depuration had no effect on the occurrence of *Cryptosporidium* oocysts in the molluscs (Gómez-Couso et al. 2003a; Freire-Santos et al. 2000), and that shellfish already harbouring *Cryptosporidium* oocysts may spread their contamination further to other shellfish at depuration plants (Gómez-Couso et al. 2003b). However, as noted by Sunnotel et al. (2007), the water circulated during depuration is UV-treated, and this decreases the viability of *Cryptosporidium* oocysts substantially; thus oocysts that contaminate further may not be of public health significance. It is those oocysts that are not purged, and that remain within the original shellfish, protected from the UV irradiation, that may pose a greater threat to the consumer.

It should be noted that as well as variation between shellfish species (e.g. Nappier et al. (2010) noted much lower depuration rates for *Cryptosporidium* in different oyster species), the success of depuration also seems to be dependent on environmental conditions such as temperature and salinity (Graczyk et al. 2006). In Sumino oysters that were experimentally exposed to *Cryptosporidium* oocysts, the depuration rate was considered to be low (oocysts still detected in the oysters over 30 days after inoculation of the water), with the most inefficient depuration observed in oysters maintained in water of low or moderate salinity (Graczyk et al. 2006). As the method used in this experiment was intended only for the detection of viable oocysts (combined IFAT and FISH assay), it was reported that even after 4 weeks post-exposure, viable *Cryptosporidium* oocysts remained in these oysters.

It is possible that if depuration water is not of an adequate standard, previously uncontaminated shellfish may then become contaminated; in a study from Brazil, although the shellfish investigated were not found to be harbouring *Cryptosporidium* oocysts, the depuration water was found to be contaminated (Leal Diego et al. 2013). However, the concentration of oocysts is not provided, and it is not clear if the contamination of the depuration water with *Cryptosporidium* originated from the shellfish, or whether the water was already contaminated prior to contact with the shellfish and might actually be a source of introduction of contamination to the shellfish.

5.5.3 Wash Water

Processing procedures post-harvest of crops, or in preparation of any other food stuff for sale, may also involve contact with water, particularly washing procedures. Such processing procedures may also result in *Cryptosporidium* oocysts being spread from contaminated food to non-contaminated food, or throughout a batch with point contamination, or, if water is used that is already contaminated with *Cryptosporidium* oocysts, then the contamination may be introduced. The extent to which such contamination is spread, or becomes a potential risk to public health, depends upon a variety of factors including the concentration of oocysts in the water, their species, their viability and infectivity, the specific contact and use of the water, and any post-contamination procedures. A study in Ireland investigated the occurrence of *Cryptosporidium* oocysts in water used to wash beef carcasses in an abattoir over the course of a year (Duffy et al. 2003); although carcass contamination was not detected, the study demonstrated that *Cryptosporidium* spp. can enter a beef abattoir in the water supply and is a potential source of carcass contamination.

In the extensive outbreak of waterborne cryptosporidiosis in Östersund, Sweden, in 2010, a manufacturer of preserved meat products became concerned that the water used in the production process might have resulted in contamination of the hams that were intended for consumption without further cooking. In particular the water used for rinsing for 3–4 h prior to smoking and drying (3,300 L per production batch, in which each batch resulted in a finished product weight of just over 1,000 kg) was considered to be a potential source of contamination, given that the water supply was known to be contaminated and had resulted in an extensive waterborne outbreak (Robertson and Huang 2012). However, despite the clear possibility for contamination under these circumstances, only one putative oocyst was detected on the meat that had been potentially exposed to contaminated water, and no ill effect on consumers was observed; whether this is because the numbers of oocysts in the rinse water were low, because the oocysts had failed to adhere to the hams, because they were inactivated and disrupted during smoking and drying, or for other reasons cannot be determined.

Another place in food production where contaminated wash water may result in a food product being contaminated is in the cleaning of the udders of cows before milking, as well as in the cleaning of the equipment used for milking. A study in Colombia investigated the water used for these purposes at 20 farms and found considerable contamination with *Cryptosporidium* oocysts (Rodríguez et al. 2012); these data demonstrate that water used in many of the processes along the farm-to-fork food production chain should be of potable quality.

5.5.4 Other Water

Dilution of juices and other beverages with contaminated water is another possible route for food contamination and, subsequently, transmission (Friedman et al. 1997).

In the fruit juices that were reported to be contaminated with *Cryptosporidium* oocysts in Egypt (Mossallam 2010), the author speculates that the source of the juice contamination may be the water that was added to the juices, rather than the fruit used for making the juice. This speculation arose because the juice that was subject to the least dilution with water (orange juice) had the lowest number of positive samples. However, water that was used for dilution of the juice was not investigated for contamination and insufficient data are provided to reach a definitive conclusion. In the Thai frozen food industry, the potential for contamination of food with protozoa, including *Cryptosporidium*, via the use of contaminated water has been investigated (Sutthikornchai et al. 2005); although treated water was not found to be contaminated, 35 % of untreated water samples contained *Cryptosporidium* oocysts, demonstrating the importance of using water of potable quality in the food industry.

Further research on the potential for water to act as a source of contamination of food is lacking. It might be noted that in major waterborne outbreaks of cryptosporidiosis, contamination of food by use of the contaminated water in the food industry, has not, to date, been proven to be problematic. However, it is possible that infections may not be traced against a background of elevated infection. Also, in industrialised countries at least, food industries often have their own barriers in place (for example, in-line UV disinfection) to ensure that potentially contaminated water does not come into contact with vulnerable processes within the industry should municipal treatment fail or be inadequate.

Chapter 6
Inactivation or Decontamination Procedures

Cryptosporidium oocysts are known to be extremely robust. They can survive for extended periods in the damp, cool conditions in which fresh produce or shellfish (likely vehicles for the transmission of *Cryptosporidium* oocysts) are generally stored, and can also survive harsher conditions such as contact with chlorine. However, they are susceptible to desiccation, heating and freeze-thawing, as well as some types of irradiation, including ultra-violet, and some chemical treatments. This resilience of *Cryptosporidium* oocysts to environmental pressures means that for food products that are intended for eating with minimal processing (no cooking) in order to retain their sensory qualities, such as taste and texture, inactivation or decontamination is probably the wrong approach; it is preferable to avoid the original contamination.

As *Cryptosporidium* oocysts as contaminants of food products must necessarily originate from either a human or animal source, the most effective means of controlling such contamination from occurring on fresh produce is application of Good Agricultural Practice (GAP) during primary production, Good Manufacturing Practice during processing and Good Hygienic Practice before consumption (Dawson 2005). GAP includes using clean water for irrigation, fertiliser/pesticide application and washing, ensuring that domestic animals do not graze or contaminate horticultural areas, and taking precautions to ensure that wild animals do not have access to these growing areas. As well as domestic animals, especially cattle and sheep, being well known as bearers of *Cryptosporidium* infection, *Cryptosporidium* oocysts have also been detected in the faeces of wild animals with access to rivers bordering agricultural areas and could pose a potential threat to human health via crop contamination (Castro-Hermida et al. 2011).

Pretreatments of wastewater prior to use for irrigation may reduce the quantity of viable pathogens, including *Cryptosporidium* oocysts, which might otherwise contaminate fresh produce; potentially suitable treatments for wastewater disinfection that may decrease the viability of oocyst include UV irradiation and ozone-treatment (Kalisvaart 2004; Orta de Velásquez et al. 2006). Membrane ultrafiltration using a submerged hollow-fibre system has also been demonstrated to be suitable for

L.J. Robertson, *Cryptosporidium as a Foodborne Pathogen*, SpringerBriefs in Food, Health, and Nutrition, DOI 10.1007/978-1-4614-9378-5_6, © Lucy J. Robertson 2014

treating wastewater prior to using it for irrigation in order to remove protozoa, including *Cryptosporidium* (Lonigro et al. 2006).

During processing procedures (e.g. washing, chopping, packaging), water of potable standard should also be used (Sutthikornchai et al. 2005), and if the wash-water is reused then the disinfection procedures should be effective at inactivating *Cryptosporidium* oocysts, as well as less resilient bacterial contaminants. It should be noted that the organic load in wash-water becomes very high (Rosenblum et al. 2012), and thus disinfectant doses that may inactivate *Cryptosporidium* oocysts in clean water may be less effective in wash-water. It should also be noted that washing procedures in themselves may be inefficient at removing *Cryptosporidium* oocysts from fresh produce; studies have demonstrated that oocysts may internalise themselves in the stomata of leafy vegetables such as spinach (Macarisin et al. 2010a), and that they may attach themselves in crevases in the surfaces of fresh produce. In addition, they have been shown to adhere strongly onto the surfaces of leaves or fruit skins (Macarisin et al. 2010b).

If contamination cannot be avoided, and elimination by washing or other procedures is inefficient, then inactivation is an additional approach for ensuring the microbiological safety of food. Due to the association of cryptosporidiosis with waterborne transmission, and the relatively high frequency of communitywide waterborne outbreaks, the bulk of research concerned with inactivation of *Cryptosporidium* oocysts has been concerned with approaches suitable for the water industry, and the supply of potable, microbiologically safe drinking water. However, survival of contaminating oocysts on different food products, and possible approaches to inactivation, has also been explored in a number of studies (see summary of some studies in Table 6.1, with greater detail provided in the following text).

When investigating the efficacy of inactivation methods on *Cryptosporidium*, it is important to have a satisfactory method by which the viability, infectivity and die-off of the oocysts can be determined (Robertson and Gjerde 2007); it is important to be aware that differences in reported inactivation rates for particular stresses could be merely artefacts associated with the different methods used by different researchers rather than actual variations in effect on the parasites themselves (King and Monis 2007). Unlike for bacteria, the cultivation of *Cryptosporidium* in vitro is complicated; although oocysts can be hatched satisfactorily in vitro, and live sporozoites seen by microscopy, this does not determine whether the sporozoites are able to invade and establish in cells. Determining infectivity using animal models (for example, mice) is possible and is considered the gold standard, but, aside from ethical issues, this method is expensive, labour-intensive and time-consuming, and not all species of *Cryptosporidium* (including *C. hominis*) are infective to mice. Cell culture methods are considered practically as good as mouse infectivity models for determining the infectivity of *C. parvum* (Rochelle et al. 2002; Garvey et al. 2010), incorporating both oocyst viability and sporozoite infectivity, with excystation in vitro followed by invasion of cultured cell monolayers. Furthermore, unlike mouse infection models, cell culture methods are also applicable to *C. hominis*. However, high levels of variability suggest that only relatively large differences in infectivity can be safely discerned by using cell culture methods.

Table 6.1 Summary of results of investigations on inactivation of *Cryptosporidium* oocysts on different food products[a]

Food type	Inactivation method	Method for assessing viability	Result	Reference
Fresh produce				
Basil and lettuce leaves	Gaseous chlorine dioxide	Cell culture methods	Effective	Ortega et al. (2008)
Various raw produce	Sodium dichloroisocyanurate	Vital dye inclusion and infectivity bioassay	Partially effective	El Zawawy et al. (2010)
Green peppers	Chlorination	Vital dye inclusion	Ineffective	Duhain et al. (2012)
	Blanching		Effective	
	Blast freezing		Partially effective	
	Microwaving		Effective	
Beverages				
Milk	Pasteurisation	Mouse infectivity	Effective	Harp et al. (1996)
Apple juice, orange juice	High hydrostatic pressure	In vitro excystation and cell culture	Effective	Slifko et al. (2000)
Apple juice (cider)	Flash pasteurisation	Cell culture methods	Effective	Deng and Cliver (2001)
Apple juice (cider), orange juice, grape juice	Organic acids and hydrogen peroxide	Cell culture methods	Effective	Kniel et al. (2003)
Apple juice (cider)	UV irradiation	Mouse infectivity assays	Effective	Hanes et al. (2002)
Shellfish				
Oysters	High hydrostatic pressure	Mouse infectivity	Partially effective	Collins et al. (2005a)
Oysters	E-beam irradiation	Mouse infectivity	Effective	Collins et al. (2005b)
Oysters	Microwave energy		Ineffective	
Blue mussels	Steaming	Mouse infectivity	Ineffective	Gómez-Couso et al. (2006c)
Oysters	UV depuration	Vital dye inclusion	Partially effective	Sunnotel et al. (2007)
Meat products				
Lean and fat beef trimmings	Blast freezing and tempering	Vital dye inclusion	Partially effective	McEvoy et al. (2004)
Beef muscle	Hot water wash of carcasses and standard thermal treatments	Cell culture assay	Effective	Moriarty et al. (2005)

[a]Further details are provided in text; investigations of potential food sanitizers in which investigations did not include inoculation into food products are not included in the table, but are described in the text

Various methods have been used for discerning infection of cell culture mono-layers, including IFAT, PCR targeting *Cryptosporidium* sp.-specific DNA and reverse transcriptase PCR (RT-PCR); although the PCR assay has been demon-strated to have the highest sensitivity, it also produces most false positives with mock-infected cells and inactivated oocysts, while IFAT tends not to result in false positives and is therefore considered most suitable for routine and sensitive detec-tion (Johnson et al. 2012).

The use of fluorescent in situ hybridisation (FISH) has been quite widely used for determining viability of *Cryptosporidium* oocysts, particularly in shellfish, and thus may have application in determining the efficacy of disinfection methods at inactivating them. However, this method, which is based upon the theory that there is rapid rRNA breakdown following cell death and thus that probes targeting spe-cific rRNA sequences should label only potentially infective or recently inactivated oocysts, has provided some confusing results; comparative experiments demon-strated only modest agreement of FISH with infectivity and cell culture assays (Jenkins et al. 2003), and gamma-irradiated (dead) oocysts have been reported to give high FISH signals (Robertson and Gjerde 2007).

A more promising method for assessing *Cryptosporidium* oocyst viability, and thus for evaluating the effects of disinfection procedures, involves the detection of mRNA transcripts found only in viable oocysts; targets used include mRNA tran-scripts from heat shock protein synthesis, β-tubulin and amyloglucosidase (Robertson and Gjerde 2007). However, the use of these methods for assessing die-off of *Cryptosporidium* oocysts in food products is lacking.

At a more basic level, evaluation of morphological criteria (as observed by microscopy, preferably with DIC optics) along with inclusion or exclusion vital dyes provides an estimate of oocyst viability, but such methods are probably too imprecise for providing an industry standard and also frequently overestimate via-bility, and thus underestimate the efficacy of a treatment that is being investigated (Erickson and Ortega 2006).

Thus, in comparing investigations of potential treatments for inactivating *Cryptosporidium* oocysts on food, conducted by different research groups, using different isolates, sometimes different species, of *Cryptosporidium*, and employing different methods for assessing viability, these potential confounders must also be considered (Dawson et al. 2004).

During many food manufacturing processes, elimination or inactivation of microbial pathogens on foods is achieved by a variety of methods, including (most commonly) heat and chemical disinfection, and also irradiation or high pressure. Although the most commonly used sanitizer is traditionally chlorine (usually applied in the form of hypochlorous acid (HOCl), which is considered to be most effective), human health and environmental concerns (production of potentially car-cinogenic by-products such as trihalomethanes and haloacetic acids and generation of wastewater with high levels of biological oxygen demand) have led some European authorities to prohibit the use of chlorine in organic produce washing. For example, several European countries, including Germany, Denmark, Holland and France, have banned the use of chlorine in washing organic produce, and alternatives, such as ozonisation or neutral electrolysed water, have been explored (Rosenblum et al. 2012; Abadias et al. 2008). Nevertheless, chlorination remains the most commonly used commercial sanitising agent, with concentrations used for food application ranging from 50 to 200 ppm. For fresh produce, the most common application is 100 ppm hypochlorite; at pH 6.8–7.1, this dose yields 30–40 ppm free chlorine, depending upon organic load, for a contact time of 2 min at 4 °C. Although exposure of *Cryptosporidium* oocysts to 1.3 ppm of chlorine dioxide yielded 90 % inactivation after 1 h, while 80 ppm of chlorine and 80 ppm of monochloramine required approximately 90 min for 90 % inactivation (Korich et al. 1990), the efficacy of standard chlorination at inactivating *Cryptosporidium* oocysts on fresh produce is dependent on a range of factors, including contact time. In particular, for some more delicate produce, such as soft fruit (e.g. strawberries and raspberries), a quick spray with, or a brief (10 s) immersion in, 15–20 ppm free chlorine is used—depending on a range of factors, this may provide insufficient contact time for effective inactivation of *Cryptosporidium* oocysts. A challenge test in which strips of green pepper inoculated with *C. parvum* oocysts were exposed to chlorination regimes (100 and 200 ppm) for 40 s and using vital dye (propidium iodide) inclusion as a measure of viability (Duhain et al. 2012) demonstrated negligible effect on oocyst viability. Other treatments investigated in this study included blanching (immersion in water at 96 °C for 3 min, then cooling for 3 min), blast freezing (−20 °C for 3 min, then thawing at 4 °C for 3 min), microwave heating in a domestic microwave oven (850 W, 2,450 MHz for 5 min) and combinations of chlorination with blast freezing and microwave treatment (Duhain et al. 2012). These studies demonstrated that although blanching was effective at killing oocysts, blast freezing was not, and it was not improved by an initial chlorination step. However, microwave treatment was effective, and its efficacy was improved by a chlorination step. Microwave treatment of *Cryptosporidium* oocysts for as little as 20 s (with temperatures reaching a minimum of 80 °C) has previously been demonstrated to inactivate *C. parvum* oocysts, with lack of viability demonstrated both by vital dye inclusion and exclusion and also neonate animal infectivity assay (Ortega and Liao 2006).

The efficacy of sodium dichloroisocyanurate (NaDCC) at inactivating various intestinal protozoa, including *Cryptosporidium* oocysts, on raw vegetables and fruits has been investigated (El Zawawy et al. 2010). Although NaDCC was more effective against other species of protozoa (including *Giardia*, *Entamoeba* and Microsporidia), the authors suggest that it may be suitable for use at the household and restaurant level, as well as in catering and fresh produce industry, mentioning its convenience in dry tablet format, and also its cheapness. Other disinfectants that may be appropriate for inactivating *Cryptosporidium* oocysts contaminating fresh produce and have been investigated experimentally include gaseous chlorine dioxide (tested on oocysts inoculated onto basil and lettuce leaves) (Ortega et al. 2008), and a combination of levulinic acid and sodium dodecyl sulphate (SDS) that was investigated on *Cryptosporidium* oocysts in suspension (Ortega et al. 2011) using cell culture to assess viability. Investigation of the latter disinfection technique was based upon the over 6 log reduction of bacterial pathogens (*Salmonella* and *E. coli* O157) on lettuce (Zhao et al. 2009). Although chlorine dioxide affected the viability of *Cryptosporidium*, as determined by cell culture methods (Ortega et al. 2008), levulinic acid with SDS had no effect, even at the highest concentrations used and with exposure up to 2 h (Ortega et al. 2011). The authors point out that this is unfortunate, as due to its non-toxic nature this disinfectant regime would have been ideal for use in the food industry. Other technologies that may be useful for using in the fresh produce industry for inactivating *Cryptosporidium* cysts include irradiation (UV, cobalt-60), high pressure processing, and ozonisation. It should be noted that the use of sequential inactivation treatments might optimise existing treatments through synergistic effects (Erickson and Ortega 2006).

The associations of cryptosporidiosis outbreaks with drinks, particularly apple cider and milk, mean that there has been some investigations on appropriate disinfection approaches for beverages, particularly as one of the apple cider associated with one of the outbreaks had been treated with ozone (Blackburn et al. 2006). One study (using vital dyes for assessing oocyst viability) demonstrated that *Cryptosporidium* does not survive well in carbonated drinks (Friedman et al. 1997). Loss of viability was also recorded in beer (three types, one without alcohol), and this was considered to be due to a low pH, and pH was also thought to contribute to the loss of viability in carbonated drinks. However, over 50 % of oocysts were considered to survive in orange juice, which was of a similarly low pH (Friedman et al. 1997). Lack of survival of *Cryptosporidium* oocysts in juices with low pH (orange juice and lemon juice) has also been reported in a study from Egypt (Mossallam 2010).

Although standard pasteurisation of milk is known to eliminate many microbial pathogens, and outbreaks of cryptosporidiosis associated with milk tend to be associated with raw milk or other dairy products, a formal study investigating whether high-temperature-short-time (HTST) pasteurisation abrogates the infectivity of *C. parvum* oocysts has also been conducted (Harp et al. 1996). Mouse infection assays were used to demonstrate that HTST (71.7 °C for 5, 10 or 15 s) successfully inactivated the oocysts. Further studies have also been conducted on other dairy products, including yogurt and ice cream (Deng and Cliver 1999), using vital dye exclusion to investigate viability of *Cryptosporidium parvum* included in their production with the milk. These studies demonstrated that viable oocysts in milk were not inactivated by the fermentation process used for yogurt making; however, the mixing and freezing process apparently resulted in complete loss of viability (Deng and Cliver 1999).

The spate of outbreaks of cryptosporidiosis associated with apple cider between 1993 and 2003 (see Table 3.1) resulted in several studies of methods for inactivating *Cryptosporidium* oocysts in this beverage, and other fruit juices. Methods investigated include high hydrostatic pressure (Slifko et al. 2000), flash pasteurisation (Deng and Cliver 2001), UV irradiation (Hanes et al. 2002), and organic acids combined with hydrogen peroxide (Kniel et al. 2003). All the methods were found to be effective, and further investigations on the probable inactivation route of the last method were investigated in further studies (Kniel et al. 2004). Either sulphydryl oxidation of proteins on oocyst surfaces or oxidation of cysteine proteases required for oocyst wall strength were suggested as possible mechanisms by which hydrogen peroxide may have a negative impact on oocyst viability.

Despite the lack of outbreaks of cryptosporidiosis associated with consumption of shellfish, the relatively common occurrence of *Cryptosporidium* oocysts in shellfish destined for human consumption, along with the fact that shellfish are frequently consumed raw or very lightly cooked, has resulted in some research on inactivation of *Cryptosporidium* in shellfish. Experimental studies have demonstrated that although a proportion of oocysts dies immediately after being ingested by shellfish (within the first 4 days), a considerable proportion remain viable and infective for a prolonged period (Tamburrini and Pozio 1999; Freire-Santos et al. 2001, 2002).

Prior to harvesting, use of UV depuration while oysters were held in industrial-scale depuration tanks has been reported to reduce the viability of experimentally added oocysts (Sunnotel et al. 2007); however, low numbers of viable oocysts were recovered from the shellfish, indicating that the parasites still represented a risk to public health. Another study that investigated the use of UV depuration, according to EU/Portuguese procedures, for cleansing a variety of species of shellfish (two species of clam, cupped oysters, blue mussels and cockles), investigated only removal rather than survival (da Fonseca et al. 2006). Their results demonstrated that oocysts can persist in bivalves after 24 h of treatment.

Post-harvesting, high pressure processing (maximum of 555 MPa for 180 s) had a significant effect on mouse infectivity of *Cryptosporidium*, although some mice (4 out of 20 at maximum dose) still became infected, and a small effect on the colour of the oyster meat (Collins et al. 2005a). Similarly, e-beam irradiation and microwave energy has been investigated as a means of inactivating *Cryptosporidium* oocysts in oysters (Collins et al. 2005b). Although a dose of 2 kGy completely eliminated *C. parvum* infectivity, as demonstrated by mouse infectivity, and without affecting the visual appearance of the oysters, microwave energy exposure (maximum of 62.5 °C for 3 s) not only was ineffective at abrogating infectivity of the oocysts but also resulted in significant alterations in the texture and colour of the oyster meat. It is worth also noting that at the preparation and consumer point, steaming of mussels experimentally contaminated with *Cryptosporidium* oocysts was found to be ineffective at killing them, as demonstrated by mouse infectivity assays; only mussels that had been steamed until their shells opened were used in this study (Gómez-Couso et al. 2006b).

Although only one small outbreak of cryptosporidiosis has been associated with meat (Yoshida et al. 2007; see Table 3.1), some experiments have been conducted to investigate the effects of standard beef preparation processes following slaughter on the viability of *Cryptosporidium* oocysts. As *Cryptosporidium* infections are relatively common in cattle, it is possible that contamination of meat with oocysts may occur at the slaughterhouse. Commercial blast freezing and tempering processes were demonstrated to have a significant effect on oocyst viability (McEvoy et al. 2004), and oocysts were entirely inactivated by standard thermal treatments, including washing with hot water (Moriarty et al. 2005).

Chapter 7
Risk Assessment and Regulations

The principal legislation that is in place to control the spread of pathogens such as *Cryptosporidium*, and similar protozoan parasites such as *Giardia*, in food within the manufacturing, processing, distribution, catering and retail sectors includes local Food Safety Acts or Regulations. The international impact of such Acts or Regulations is generally insignificant, as they tend to be directed towards specific local problems. The emergence of novel foodborne pathogens provides a challenge to such regulations. Nevertheless, the foremost food safety management system for many years (originally devised to ensure that the foods consumed by astronauts were safe) has been Hazard Analysis and Critical Control Point (HACCP), supported by Good Hygienic Practice (GHP) and Good Manufacturing Practice (GMP). The intention of HACCP is to provide a systematic, cost-effective and efficacious approach for risk management and prevention; HACCP is of international application.

In addition, responsible authorities set public health targets that must be included by those involved in the provision of food. For example, the World Trade Organization introduced the concept of appropriate level of protection (ALOP) as a public health target, and other concepts have been introduced to enable these targets to be translated into meaningful, tangible objectives for the food industry. Among these concepts are the following: Food Safety Objectives (FSOs), Performance Objectives (POs) and Performance Criteria (PC), which were proposed by the International Commission on Microbiological Specifications for Foods (ICMSF) and adopted by the Codex Alimentarius Food Hygiene Committee. How these FSOs can be extrapolated from or to ALOPs has yet to be established, but it is envisaged that FSOs will provide a functional link between risk assessment (RA) and risk management, with HACCP acknowledged to be the principle tool available for use in the food industry.

While HACCP enables food producers to identify hazards and measures for their control, and the determination of critical control points (CCP) along the farm/fjørd-to-fork continuum, it differs from risk assessment. Risk assessment, comprising hazard identification, hazard characterisation, exposure assessment and risk characterisation, is a complementary tool to HACCP. HACCP enables consideration of

L.J. Robertson, *Cryptosporidium as a Foodborne Pathogen*, SpringerBriefs in Food, Health, and Nutrition, DOI 10.1007/978-1-4614-9378-5_7, © Lucy J. Robertson 2014

multiple hazards for a single product in a particular facility, while risk assessment traditionally focuses on single pathogen–food combinations. Nevertheless, despite this difference in emphasis, for both HACCP and risk assessment, the key objective is risk mitigation. The HACCP system is based on seven principles, each of which has a specific aim or seeks to answer a key question (see Robertson et al. 2014):

- Hazard analysis (key aim and/or questions: what are the food safety hazards? how can they be prevented?)
- Identification of CCP (key aim and/or questions: at which points in the food chain can controls be applied resulting in the prevention, elimination or reduction of hazards to an acceptable level?)
- Specification of criteria to ensure control (key aim and/or questions: what levels of the hazard are acceptable in the food products? how can these levels be determined and identified?)
- Monitoring of critical CCP (key aim and/or questions: establishment of a system by which CCP can be monitored; what is the frequency and method for each monitoring procedure?)
- Implementation of corrective action whenever monitoring indicates that criteria are not being met (key aim and/or questions: establishment of appropriate corrective actions when a CCP is not under control; which corrective actions should be implemented if a critical limit is exceeded?)
- Validation and verification that the system is functioning as planned (key aim and/or questions: establishment of procedures for verification to confirm that HACCP is working effectively)
- Documentation for all procedures and records appropriate to these principles and their applications (key aim and/or questions: maintenance of HACCP plans; documentation of monitoring; documentation of verification activities and handling of processing deviations)

These principles have similarities, and differences, to the four established stages in risk assessment:

- Hazard identification: identification of agents that may cause adverse health effects and that may be occurring in a particular food or food group.
- Exposure assessment: qualitative and/or quantitative evaluation of the likely intake of the relevant agents via food (also exposures from other sources if relevant).
- Hazard characterisation: qualitative and/or quantitative evaluation of the nature of the adverse health effects. For microbiological risk assessment, the concerns relate directly to microorganisms (and/or, where relevant, their toxins). Infectious dose is also of relevance.
- Risk characterisation: qualitative and/or quantitative estimation, including attendant uncertainties, of the probability of occurrence and severity of known or potential adverse health effects in a given population based on the three previous steps.

Quantitative microbiological risk assessment (QMRA) is based on risk assessment, but has the fundamental aims of protecting consumers from microbial risks, enabling decision-making on food safety issues and assisting relevant authorities to meet public health goals by providing numerical limits and targets (within defined ranges). By focussing on risks that are associated with single hazards and product groups, it is possible to identify products of greatest concern to public health and those aspects of their processing, handling and supply that are likely to have most impact on risk. Thus, there is a clear overlap with, and support for, HACCP. While risk assessment in general, and QMRA in particular, can enhance HACCP by aiding in the identification of 'design' CCPs, detailed considerations of specific facilities or locations are not a part of risk assessments with a more general basis.

QMRAs are very data demanding, and, for protozoan parasites including *Cryptosporidium*, obtaining the data that are necessary to develop a robust QMRA is challenging. However, by reviewing the available published data and assessing it critically, it is possible to identify the gaps that must be filled before the QMRA can be conducted. Data can be acquired from a range of different sources, including retrieval from publications, by expert elicitations, by conducting relevant studies or by developing models. More recently developed QMRAs tend to be stochastic in nature, with probability distributions replacing single values, thereby providing a more accurate reflection of the uncertainty associated with inputs and derived parameters. It is obvious that as the relevance and accuracy of the data fed into a QMRA increase, the uncertainty in the risk estimates decreases correspondingly, and thus, the guidance derived is more useful for decision-making. Documentation and, thereby, transparency are also key factors. Microbial risk assessment, particularly with application to assessing the risk of infection from drinking water, first became a commonly used tool around 20–30 years ago, and QMRA is frequently utilised in the water industry, with reference to *Cryptosporidium* as well as other pathogens. Within the water industry, QMRA is considered to be essential in the construction of Water Safety Plans (e.g. Smeets et al. 2010), addressing such questions as: 'how safe is the water?' 'how much does the safety vary?' and 'how certain is the estimated of safety?' Acknowledgement of the importance of QMRA in providing risk managers in water utilities with answers to such questions has led to the development of user-friendly, online tools specifically for water industries. In such tools, the necessary data can be loaded by an operator who has no requirement for special experience in QMRA modelling, and a risk outcome associated with a particular pathogen, including *Cryptosporidium*, for consumption of drinking water from a particular source will be provided (Schijven et al. 2011). Such tools have not been widely implemented for irrigation water, although the principle would be the same, and quantitative risk assessment has been published that attempts to estimate the numbers of people that would be affected by fresh produce irrigated with water contaminated with protozoan parasites, including *Cryptosporidium* (Mota et al. 2009). In this particular study (Mota et al. 2009), input into the QMRA included the following:

- The number of *Cryptosporidium* oocysts per 100 L of irrigation water, with 58 surface water samples collected from widely dispersed points in one of the largest horticultural production areas of Mexico, and including samples taken from rivers, irrigation channels and drainage channels

- Results ranged from 17 to 200 oocysts per 100 L with a geometric mean of 32.94 oocysts per 100 L.

- The recovery efficiency of the detection method

  - 15 %.

- The volume of irrigation water estimated to be retained on fresh produce

  - This varied according to produce type and was estimated to be 0.0036 mL/g for tomatoes, bell peppers and cucumbers and 0.108 mL/g for lettuce (based on data from Shuval et al. 1997).

- The daily consumption of the different produce types by an adult in the USA

  - This varied according to produce type and was estimated to be 13.0 g (tomatoes), 4.3 g (bell peppers), 3.3 g (cucumbers) and 6.2 g (lettuce) (based on information from an online food consumption database run by the US Department of Agriculture).

For the purposes of the QMRA, a worst-case scenario was assumed, with 100 % transfer of the *Cryptosporidium* oocysts from the irrigation water to the produce, and all the oocysts being infectious to humans. Use of a 'worst-case scenario' in such risk assessments, assuming, for example, raw consumption of produce, or overhead irrigation, is not unusual, particularly when data are limited (Hamilton et al. 2006b). From this input, the estimated annual risks of infection (assuming 120 days exposure per year) with the *Cryptosporidium* oocysts on the produce were found to range from 9.0×10^{-6}, associated with bell peppers contaminated via irrigation water with the lowest concentration of *Cryptosporidium* oocysts (17 oocysts per 100 L), up to 4.48×10^{-3}, associated with lettuce contaminated via irrigation water with the highest concentration of *Cryptosporidium* oocysts (200 oocysts per 100 L) (Mota et al. 2009).

Despite limitations within a process such as this, and the requirement for making unsubstantiated assumptions, it nevertheless provides data that are useful for considering appropriate mitigation strategies/intervention practices or guidelines for pathogen reduction requirements. Other QMRAs have looked at specific processes. For example, the use of wastewater or sewage sludge, including for fertilising vegetable crops, and the potential for spread of pathogens, including *Cryptosporidium*, has been considered using QMRA and HACCP (Westrell et al. 2004), with the worst-case situation with the largest number of infections considered to arise when vegetables are fertilised with sludge and eaten raw (as this is illegal in Sweden, where the study was conducted, this is an improbable scenario, but may very well be probable in other countries).

In another study, an assessment tool requiring specific information on weather, topography, vegetation, land management practices, etc. was used to investigate those activities that regulate the risk of surface water contamination (potentially used for irrigation water) with *Cryptosporidium* oocysts from agricultural areas at two sites in Ireland (Tang et al. 2011). A sensitivity analysis used in this study suggested that temperature was the most important variable regulating oocyst transport into study catchments and that the timing of manure application with respect to water runoff events was critical (Tang et al. 2011). Interestingly, the data indicated that while grazing management had little influence on predicted oocyst transport, fields that were fertilised with manure were critical sources for contamination in the catchments being studied. In this context, it should also be noted that in a study in which the die off of *Cryptosporidium* oocysts in farmyard manure and slurries spread over pasture was investigated, a 1 log reduction was only achieved after 8–31 days, with viability assessed using vital dye exclusion (Hutchison et al. 2005).

Another QMRA sought to investigate the risk of infection from root crops grown on land to which treated sewage sludge had been applied according to UK regulations (Gale 2005). In this QMRA, event trees were used to model partitioning of pathogens from raw sewage to sewage sludge using published data, then to incorporate factors such as decay in soil, dilution in soil, remaining in soil at harvest (assumed that root crops comprise 2 % soil (by weight) at harvest) and removal during washing. These data, which assumed a linear decay rate in soil, indicated that a 12-month harvest interval (period of 12 months between application of sewage sludge to soil and harvesting of the crop) resulted in a prediction of a single infection with *C. parvum* in the UK, via this infection route, every 45 years (Gale 2005). The authors conclude that the risk to humans from consumption of vegetable crops is remote and that any lapses in operational efficiency of sludge treatment would be compensated for by the stipulated harvest intervals. It should be noted that while root crops have not yet been associated with *Cryptosporidium* infections, this study does not apply to leaf crops or fruits.

In such QMRAs that seek to address a specific question, it is possible that some contamination routes may be overlooked; thus, it is important that for any particular food item, the assessor has a clear overview of potential routes of contamination, or how, for *Cryptosporidium*, oocysts might come into contact with the specific food. This is often illustrated via a flow diagram (see, e.g. Moore et al. (2007) for a figure illustrating possible route of entry and control of *Cryptosporidium* in the production of lettuce). More all-purpose diagrams, directed to a range of food products, are provided in other articles (e.g. Budu-Amoako et al. 2011; Robertson and Chalmers 2013).

In a review of foodborne illness associated with *Cryptosporidium* and *Giardia* from livestock, Budu-Amoako et al. (2011) listed a range of interventions and mitigation strategies to reduce the contamination of fresh produce. These include the following: on-farm interventions (GAP) to minimise infection in animals and further transmission of infection, watershed interventions to reduce contamination of water sources from parasites excreted from infected animals (GAP) and food processing plant interventions to prevent food contamination (GHP and GMP), with emphasis on the application of HACCP. The authors emphasise the importance of prevention of environmental contamination, particularly water sources, and mention the use of artificial drains and buffer strips as possible interventions to limit such contamination and also note the importance of proper treatment and management of manure. However, there is limited emphasis on mixed farms, where animals and food crops may both be raised, and the importance of ensuring that sufficient barriers are kept between the two types of agricultural commodities, both spatially and temporally, such that transfer from animals to plants via equipment or human transfer is minimised by ensuring that specific interventions (such as use of protective clothing or dedicated equipment) are instigated and followed. Various outbreaks of cryptosporidiosis, both foodborne (e.g. Collier et al. 2011; see table 2) and probably direct (Silverlås et al. 2012), demonstrate the importance of on-farm hygiene regarding transfer of *Cryptosporidium* oocysts from animal excretion to human ingestion.

Chapter 8
Future Challenges

Although we continually acquire more information and improve those technologies that can be used to combat foodborne transmission of cryptosporidiosis, a changing globe poses new challenges. Improved refrigeration for transport of fresh produce means that we now import or export our food as never before, producing complicated trade routes, more handling and greater possibilities for contamination, as well as potential difficulties for trace-back should contamination events occur; globalisation of foodborne parasites, including *Cryptosporidium*, has been a subject of increasing concern (Robertson et al. 2014). While improved food traceability systems have the potential to be used for some produce (e.g. radio-frequency identification; Kumar et al. 2009), the application of such technologies for third countries and small-scale producers is challenging and implementation will probably be driven by the perceived cost–benefits as well as by consumer demand (Robertson and Chalmers 2013). Globalisation applies not only to food products but also to the people who handle food along the farm-to-fork continuum and to the people who consume products—travel and tourism continue to flourish despite global economic downturns—and seasonal tasks, such as crop picking, often involve the use of an itinerant workforce. People travel with their intestinal parasites, and if the possibility arises for contamination of food products, then other people, perhaps in otherwise naïve populations, may also become exposed.

Globalisation may also be considered to apply to human behaviours and habits, and this may be important with respect to foodborne transmission of a range of parasites, including *Cryptosporidium* (Robertson et al. 2014). This can include trends in which efforts are made to increase the consumption of vegetables by children in order to attempt to combat increases in chronic diseases such as cardiovascular complaints (e.g. Wolfenden et al. 2012); trends of eating more 'exotic' fruits and vegetables, often imported to industrialised nations from countries where contamination may be more likely to occur due to infrastructure problems, as well as a higher endemicity of infection in the local population (Unnevehr 2000); and trends of urban fast-food cafes including salads as part of their menus in areas of the world where fresh, uncooked vegetables are not part of the normal diet (Amoah et al. 2007).

L.J. Robertson, *Cryptosporidium as a Foodborne Pathogen*, SpringerBriefs in Food, Health, and Nutrition, DOI 10.1007/978-1-4614-9378-5_8, © Lucy J. Robertson 2014

Furthermore, the rise of interest in organic farming globally may have an impact on foodborne transmission of *Cryptosporidium*; although perceived as healthier and safer, particularly with respect to agrochemical residues, relevant scientific evidence is scarce and the available evidence regarding contamination with microbial pathogens is so limited that it is not possible to make generalised statements (Magkos et al. 2006). While the absence of data means that it is impossible to weigh the risks, it is obvious that 'organic' does not automatically equal 'safe'. However, globalisation need not be all bad; by facilitating the international sharing of methods and information, globalisation might also provide approaches for controlling the entry of parasites, such as *Cryptosporidium*, into the global food chain (Robertson et al. 2014).

Climate change is another global challenge that may impact on the foodborne transmission of microbial pathogens, including *Cryptosporidium* (Miraglia et al. 2009). In general, climate change models predict higher temperatures, heat waves, excessive precipitation, storm surges, droughts and floods, many of which are likely to affect the exposure pathways of food products to contamination. In order to address this challenge, a computational tool has been developed (Schijven et al. 2013) for estimating climate change-associated relative infection risks for different pathogens, including *Cryptosporidium*. The intention is that by inputting location-specific data under current and projected climate change conditions, the tool can be used to guide intervention strategies to limit or reduce the risk of infection (Schijven et al. 2013).

Human population factors may also present a challenge with respect to cryptosporidiosis. An increasing population of elderly people or patients with immuno-suppressive diseases may result in populations that are more susceptible to foodborne cryptosporidiosis and have more severe symptoms when infected (Nuñez and Robertson 2012). In addition, as human populations expand, greater intrusions and interactions occur between wildlife, domestic animals and people; it is at such interfaces that new strains with different host specificities, pathogenicities and sensitivities might arise, and that may pose a greater threat to foodborne transmission.

Chapter 9
Conclusions

Cryptosporidium is often considered as a parasitic infection of relatively minor severity, usually associated with low mortality and morbidity, and generally self-resolving. However, clinical cryptosporidiosis in some populations nevertheless is also associated with prolonged and unpleasant symptoms, which may, in specific populations or circumstances, result in, or contribute to, death. This is particularly because of the lack of effective treatments. The recent identification of *Cryptosporidium* as being one of the four main aetiological agents that are associated with serious childhood diarrhoea in developing countries is particularly important (Kotloff et al. 2013). In developed countries, foodborne cryptosporidiosis can still have a major impact, particularly in outbreak situations; a study from the USA that sought to estimate the number of episodes of foodborne illness, hospitalisations and deaths caused by specific pathogens (Scallan et al. 2011) calculated that, based on the US population in 2006 (299 million persons), the estimated annual number of episodes of domestically acquired foodborne cryptosporidiosis was over 57,000 (with 90 % credible interval (CrI) of 12,060–166,771), while the estimated mean annual number of hospitalisations and deaths due to domestically acquired foodborne cryptosporidiosis were 210 (90 % CrI 58–518) and 4 (90 % CrI 0–19), respectively (Scallan et al. 2011). An earlier study from the UK (Adak et al. 2002) provides similar estimates and suggests that in 1995, there would have been over 2000 cases of indigenous foodborne cryptosporidiosis, 42 hospital admissions and 4 deaths due to *Cryptosporidium parvum* infection in England and Wales, while in 2000, the corresponding data were estimated to be again over 2000 cases and 39 hospital admissions, but <1 death. It seems possible that in deriving these data, the potential for foodborne cryptosporidiosis was underestimated.

In association with other insults to health, especially concomitant infections and conditions (particularly those that affect immunocompetence), or compromised nutritional status, the effects of cryptosporidiosis may be particularly severe. Cryptosporidiosis may thus be considered a relatively 'neglected disease' (Savioli et al. 2006), both in countries with less advanced infrastructures where it weakens the ability to achieve full potential and impairs development and socio-economic improvements, but also in the most wealthy countries in the world. It should not be

L.J. Robertson, *Cryptosporidium as a Foodborne Pathogen*, SpringerBriefs in Food, Health, and Nutrition, DOI 10.1007/978-1-4614-9378-5_9, © Lucy J. Robertson 2014

forgotten that outbreaks of cryptosporidiosis continue to occur in wealthy countries (e.g. UK, Sweden, USA) where they cause considerable discomfort to those affected and considerable disruption to local economies. The lack of effective treatments for the most vulnerable patient groups remains an important barrier to the control of this infection.

Although *Cryptosporidium* is known as a pathogen that is transmitted via the faecal–oral route, and it is particularly associated with waterborne transmission and waterborne outbreaks, it is less frequently associated with food as an infection vehicle. Nevertheless, many foodborne outbreaks have been documented, and the potential for foodborne infection is evident; the transmission stage is extremely resilient to environmental pressures, the infectious dose is low and *Cryptosporidium* oocysts are excreted in enormous quantities by infected hosts. Thus, one probable reason that infections with *Cryptosporidium* are relatively infrequently recognised as foodborne is not because foodborne transmission occurs seldom, but because clinical, diagnostic and epidemiological barriers hamper the ease of making the correct associations; the symptoms may start several days after the implicated food has been eaten, the clinician may fail to request the appropriate samples or request the appropriate tests, the diagnostician may fail to make the correct diagnosis and the resultant period between consumption of contaminated food and the diagnosis of cryptosporidiosis frustrates epidemiological investigation. Therefore, foodborne cryptosporidiosis probably occurs considerably more often than would be expected by direct extrapolation from the cases reported in the literature.

How food becomes contaminated, which food products are most likely to be contaminated and how *Cryptosporidium* oocysts that have contaminated food can be removed and/or inactivated are all questions about which the data available are relatively limited. Nevertheless, several surveys have been conducted, and these indicate that fresh produce (in particular), beverages, shellfish and possibly meat can act as potential food vehicles for transmission. These data, along with several outbreaks, have been driving forces for developing a standardised method for examination of food (especially fresh produce—leafy greens and small red fruit) for contamination with *Cryptosporidium* oocysts, and an ISO Method, based on elution, concentration, isolation and detection, is under development (registered in the ISO/TC34/SC9 work program with the number ISO 18744). The use of molecular methods to determine the species and subtype of *Cryptosporidium* oocysts that have been detected can be added on to the method subsequently.

While the information that is obtained using such methods is valuable for increasing our understanding of foodborne transmission of cryptosporidiosis, there are also obvious gaps in our knowledge. These include not only our understanding of the difficulty of treating cryptosporidiosis—but also we require better methods for evaluating the viability and infectivity of small numbers of oocysts, and understanding why they are tenacious at attaching to different product types, and thereby developing appropriate methods for removal or inactivation of *Cryptosporidium* oocysts along the farm to fork continuum. In the absence of effective methods, the rigorous use of HACCP and risk analysis in order to reduce contamination and to optimise the implementation of appropriate interventions that will minimise transmission risk are very important.

References

Abadias M, Usall J, Oliveira M, Alegre I, Viñas I (2008) Efficacy of neutral electrolyzed water (NEW) for reducing microbial contamination on minimally-processed vegetables. Int J Food Microbiol 123(1–2):151–158. doi:10.1016/j.ijfoodmicro.2007.12.008

Adak G, Long S, O'Brien S (2002) Trends in indigenous foodborne disease and deaths, England and Wales: 1992 to 2000. Gut 51:832–841

Adesiyun AA, Webb LA, Romain H, Kaminjolo JS (1996) Prevalence of *Salmonella*, *Listeria monocytogenes*, *Campylobacter* spp., *Yersinia enterocolitica* and *Cryptosporidium* spp. in bulk milk, cows' faeces and effluents of dairy farms in Trinidad. Rev Elev Med Vet Pays Trop 49(4):303–309

Adl SM, Simpson AG, Farmer MA, Andersen RA, Anderson OR, Barta JR, Bowser SS, Brugerolle G, Fensome RA, Fredericq S, James TY, Karpov S, Kugrens P, Krug J, Lane CE, Lewis LA, Lodge J, Lynn DH, Mann DG, McCourt RM, Mendoza L, Moestrup O, Mozley-Standridge SE, Nerad TA, Shearer CA, Smirnov AV, Spiegel FW, Taylor MF (2005) The new higher level classification of eukaryotes with emphasis on the taxonomy of protists. J Eukaryot Microbiol 52(5):399–451

Adl SM, Leander BS, Simpson AG, Archibald JM, Anderson OR, Bass D, Bowser SS, Brugerolle G, Farmer MA, Karpov S, Kolisko M, Lane CE, Lodge DJ, Mann DG, Meisterfeld R, Mendoza L, Moestrup Ø, Mozley-Standridge SE, Smirnov AV, Spiegel F (2007) Diversity, nomenclature, and taxonomy of protists. Syst Biol 56(4):684–689

Adl SM, Simpson AG, Lane CE, Lukeš J, Bass D, Bowser SS, Brown MW, Burki F, Dunthorn M, Hampl V, Heiss A, Hoppenrath M, Lara E, Le Gall L, Lynn DH, McManus H, Mitchell EA, Mozley-Stanridge SE, Parfrey LW, Pawlowski J, Rueckert S, Shadwick RS, Schoch CL, Smirnov A, Spiegel FW (2012) The revised classification of eukaryotes. J Eukaryot Microbiol 59(5):429–493

Al-Brikan FA, Salem HS, Beeching N, Hilal N (2008) Multilocus genetic analysis of *Cryptosporidium* isolates from Saudi Arabia. J Egypt Soc Parasitol 38(2):645–658

Amoah P, Drechsel P, Henseler M, Abaidoo RC (2007) Irrigated urban vegetable production in Ghana: microbiological contamination in farms and markets and associated consumer risk groups. J Water Health 5(3):455–466

Amorós I, Alonso JL, Cuesta G (2010) *Cryptosporidium* oocysts and *Giardia* cysts on salad products irrigated with contaminated water. J Food Prot 73:1138–1140

Anonymous (1996) Foodborne outbreak of diarrhoea illness associated with *Cryptosporidium parvum*, Minnesota, 1995. Morb Mortal Wkly MMWR 45:783–784

Anonymous (1997) Outbreaks of *Escherichia coli* 0157:H7 infection and cryptosporidiosis associated with drinking unpasteurized apple cider, Connecticut and New York, October 1996. Morb Mortal Wkly MMWR 46:4–8

L.J. Robertson, *Cryptosporidium as a Foodborne Pathogen*, SpringerBriefs in Food, Health, and Nutrition, DOI 10.1007/978-1-4614-9378-5, © Lucy J. Robertson 2014

Anonymous (2013) Outbreak of cryptosporidiosis in England and Scotland, May 2012. Health Prot Rep 7(12). http://www.hpa.org.uk/hpr/news/default_0413_aa_crptsprdm.htm

Armon R, Gold D, Brodsky M, Oron G (2002) Surface and subsurface irrigation with effluents of different qualities and presence of *Cryptosporidium* oocysts in soil and on crops. Water Sci Technol 46:115–122

Azami M, Moghadam DD, Salehi R, Salehi M (2007) The identification of *Cryptosporidium* species (protozoa) in Isfahan, Iran by PCR-RFLP analysis of the 18s rRNA gene. Mol Biol (Mosk) 41:851–856

Baldursson S, Karanis P (2011) Waterborne transmission of protozoan parasites: review of worldwide outbreaks—an update 2004–2010. Water Res 45(20):6603–6614. doi:10.1016/j.watres.2011.10.013

Ben Ayed L, Yang W, Widmer G, Cama V, Ortega Y, Xiao L (2012) Survey and genetic character-ization of wastewater in Tunisia for *Cryptosporidium* spp., *Giardia duodenalis*, *Enterocytozoon bieneusi*, *Cyclospora cayetanensis* and *Eimeria* spp. J Water Health 10(3):431–444. doi:10.2166/wh.2012.204

Bier JW (1991) Isolation of parasites on fruits and vegetables. Southeast Asian J Trop Med Public Health 22(Suppl):144–145

Blackburn BG, Mazurek JM, Hlavsa M, Park J, Tillapaw M, Parrish M, Salehi E, Franks W, Koch E, Smith F, Xiao L, Arrowood M, Hill V, da Silva A, Johnston S, Jones JL (2006) Cryptosporidiosis associated with ozonated apple cider. Emerg Infect Dis 12(4):684–686

Bohaychuk VM, Bradbury RW, Dimock R, Fehr M, Gensler GE, King RK, Rieve R, Romero Barrios P (2009) A microbiological survey of selected Alberta-grown fresh produce from farmers' markets in Alberta, Canada. J Food Prot 72(2):415–420

Bouzid M, Hunter PR, Chalmers RM, Tyler KM (2013) *Cryptosporidium* pathogenicity and virulence. Clin Microbiol Rev 26(1):115–134

Budu-Amoako E, Greenwood SJ, Dixon BR, Barkema HW, McClure JT (2011) Foodborne illness associated with *Cryptosporidium* and *Giardia* from livestock. J Food Prot 74(11):1944–1955. doi:10.4315/0362-028X.JFP-11-107

Calvo M, Carazo M, Arias ML, Chaves C, Monge R, Chinchilla M (2004) Prevalence of *Cyclospora* sp., *Cryptosporidium* sp, microsporidia and fecal coliform determination in fresh fruit and vegetables consumed in Costa Rica. Arch Latinoam Nutr 54(4):428–432

Cama VA, Bern C, Sulaiman IM, Gilman RH, Ticona E, Vivar A, Kawai V, Vargas D, Zhou L, Xiao L (2003) *Cryptosporidium* species and genotypes in HIV-positive patients in Lima, Peru. J Eukaryot Microbiol 50(Suppl):531–533

Cama VA, Ross JM, Crawford S, Kawai V, Chavez-Valdez R, Vargas D, Vivar A, Ticona E, Navincopa M, Williamson J, Ortega Y, Gilman RH, Bern C, Xiao L (2007) Differences in clinical manifestations among *Cryptosporidium* species and subtypes in HIV-infected persons. J Infect Dis 196(5):684–691

Cama VA, Bern C, Roberts J, Cabrera L, Sterling CR, Ortega Y, Gilman RH, Xiao L (2008) *Cryptosporidium* species and subtypes and clinical manifestations in children, Peru. Emerg Infect Dis 14(10):1567–1574

Cann KF, Thomas DR, Salmon RL, Wyn-Jones AP, Kay D (2013) Extreme water-related weather events and waterborne disease. Epidemiol Infect 141(4):671–686. doi:10.1017/S0950268812001653

Casemore DP, Jessop EG, Douce D, Jackson FB (1986) *Cryptosporidium* plus *Campylobacter*: an outbreak in a semi-rural population. J Hyg (Lond) 96(1):95–105

Casteel MJ, Sobsey MD, Mueller JP (2006) Fecal contamination of agricultural soils before and after hurricane-associated flooding in North Carolina. J Environ Sci Health A Tox Hazard Subst Environ Eng 41(2):173–184

Castro-Hermida JA, García-Presedo I, González-Warleta M, Mezo M (2011) Prevalence of *Cryptosporidium* and *Giardia* in roe deer (*Capreolus capreolus*) and wild boars (*Sus scrofa*) in Galicia (NW, Spain). Vet Parasitol 179(1–3):216–219. doi:10.1016/j.vetpar.2011.02.023

Chaidez C, Soto M, Gortares P, Mena K (2005) Occurrence of *Cryptosporidium* and *Giardia* in irrigation water and its impact on the fresh produce industry. Int J Environ Health Res 15: 339–345

Chaidez C, Soto M, Campo NC (2007) Effect of water suspended particles on the recovery of Cryptosporidium parvum from tomato surfaces. J Water Health 5(4):625–631

Chalmers RM (2012) Waterborne outbreaks of cryptosporidiosis. Ann Ist Super Sanita 48(4): 429–446

Chalmers RM, Sturdee AP, Mellors P, Nicholson V, Lawlor F, Kenny F, Timpson P (1997) *Cryptosporidium parvum* in environmental samples in the Sligo area, Republic of Ireland: a preliminary report. Lett Appl Microbiol 25(5):380–384

Chalmers RM, Elwin K, Thomas AL, Guy EC, Mason B (2009) Long-term *Cryptosporidium* typing reveals the aetiology and species-specific epidemiology of human cryptosporidiosis in England and Wales, 2000 to 2003. Euro Surveill 14(2)

Chalmers RM, Elwin K, Hadfield SJ, Robinson G (2011) Sporadic human cryptosporidiosis caused by *Cryptosporidium cuniculus*, United Kingdom, 2007–2008. Emerg Infect Dis 17(3): 536–538

Chang'a JS, Robertson LJ, Mtambo MM, Mdegela RH, Løken T, Reksen O (2011) Unexpected results from large-scale cryptosporidiosis screening study in calves in Tanzania. Ann Trop Med Parasitol 105(7):513–519. doi:10.1179/2047773211Y.0000000007

Chappell CL, Okhuysen PC, Sterling CR, DuPont HL (1996) *Cryptosporidium parvum*: intensity of infection and oocyst excretion patterns in healthy volunteers. J Infect Dis 173(1): 232–236

Chappell CL, Okhuysen PC, Sterling CR, Wang C, Jakubowski W, Dupont HL (1999) Infectivity of *Cryptosporidium parvum* in healthy adults with pre-existing anti-C. Parvum serum immuno-globulin G. Am J Trop Med Hyg 60(1):157–164

Chappell CL, Okhuysen PC, Langer-Curry R, Widmer G, Akiyoshi DE, Tanriverdi S, Tzipori S (2006) *Cryptosporidium hominis*: experimental challenge of healthy adults. Am J Trop Med Hyg 75(5):851–857

Chappell CL, Okhuysen PC, Langer-Curry RC, Akiyoshi DE, Widmer G, Tzipori S (2011) *Cryptosporidium meleagridis*: infectivity in healthy adult volunteers. Am J Trop Med Hyg 85(2):238–242

Checkley W, Epstein LD, Gilman RH, Black RE, Cabrera L, Sterling CR (1998) Effects of *Cryptosporidium parvum* infection in Peruvian children: growth faltering and subsequent catch-up growth. Am J Epidemiol 148(5):497–506

Cieloszyk J, Goñi P, García A, Remacha MA, Sánchez E, Clavel A (2012) Two cases of zoonotic cryptosporidiosis in Spain by the unusual species *Cryptosporidium ubiquitum* and *Cryptosporidium felis*. Enferm Infecc Microbiol Clin 30(9):549–551

Clancy JL, Hargy TM (2008) Waterborne: drinking water. In: Fayer R, Xiao L (eds) *Cryptosporidium* and cryptosporidiosis, 2nd edn. CRC Press/IWA Publishing, Boca Raton, FL, pp 335–370

Collier SA, Smith S, Lowe A, Hawkins P, McFarland P, Salyers M, Rocco P, Bumby G, Maillard J-M, Williams C, Fleischauer A, Radke V, Roberts JM, Hightower AW, Bishop HS, Mathison BA, da Silva AJ, Carpenter J, Hayden AS, Hlavsa MC, Xiao L, Roberts VA, Brunkard J, Beach MJ, Hill V, Yoder J, Dunbar EL, Dearen T, Bopp C, Humphrys MS, Phillips G, Chang L, Meites EM (2011) Cryptosporidiosis outbreak at a summer camp—North Carolina. Morb Mortal Wkly MMWR 60:918–922

Collins MV, Flick GJ, Smith SA, Fayer R, Croonenberghs R, O'Keefe S, Lindsay DS (2005a) The effect of high-pressure processing on infectivity of *Cryptosporidium parvum* oocysts recovered from experimentally exposed Eastern oysters (*Crassostrea virginica*). J Eukaryot Microbiol 52(6):500–504

Collins MV, Flick GJ, Smith SA, Fayer R, Rubendall E, Lindsay DS (2005b) The effects of E-beam irradiation and microwave energy on Eastern Oysters (*Crassostrea virginica*) experimentally infected with *Cryptosporidium parvum*. J Eukaryot Microbiol 52(6):484–488

Conn DB, Weaver J, Tamang L, Graczyk TK (2007) Synanthropic flies as vectors of *Cryptosporidium* and *Giardia* among livestock and wildlife in a multispecies agricultural complex. Vector Borne Zoonotic Dis 7(4):643–651

Cook N, Paton CA, Wilkinson N, Nichols RA, Barker K, Smith HV (2006a) Towards standard methods for the detection of *Cryptosporidium parvum* on lettuce and raspberries. Part 1: development and optimization of methods. Int J Food Microbiol 109(3):215–221

Cook N, Paton CA, Wilkinson N, Nichols RA, Barker K, Smith HV (2006b) Towards standard methods for the detection of *Cryptosporidium parvum* on lettuce and raspberries. Part 2: validation. Int J Food Microbiol 09(3):222–228

Cook N, Nichols RAB, Wilkinson N, Paton CA, Barker K, Smith HV (2007) Development of a method for the detection of *Giardia duodenalis* on lettuce and for simultaneous analysis of salad products from the presence of *Giardia* cysts and *Cryptosporidium* oocysts. Appl Environ Microbiol 73:7388–7391

da Fonseca IP, Ramos PS, Ruano FA, Duarte AP, Costa JC, Almeida AC, Falcão ML, Fazendeiro MI (2006) Efficacy of commercial cleansing procedures in eliminating *Cryptosporidium parvum* oocysts from bivalves. J Eukaryot Microbiol 53(Suppl 1):S49–S51

Dawson D (2005) Foodborne protozoan parasites. Int J Food Microbiol 103:207–227

Dawson DJ, Samuel CM, Scrannage V, Atherton CJ (2004) Survival of *Cryptosporidium* species in environments relevant to foods and beverages. J Appl Microbiol 96(6):1222–1229

Deng MQ, Cliver DO (1999) *Cryptosporidium parvum* studies with dairy products. Int J Food Microbiol 46(2):113–121

Deng MQ, Cliver DO (2000) Comparative detection of *Cryptosporidium parvum* oocysts from apple juice. Int J Food Microbiol 54(3):155–162

Deng MQ, Cliver DO (2001) Inactivation of *Cryptosporidium parvum* oocysts in cider by flash pasteurization. J Food Prot 64(4):523–527

Deng MQ, Lam KM, Cliver DO (2000) Immunomagnetic separation of *Cryptosporidium parvum* oocysts using MACS MicroBeads and high gradient separation columns. J Microbiol Methods 40(1):11–17

Di Benedetto MA, Di Piazza F, Oliveri R, Cerame G, Valenti R, Firenze A (2006) Development of a technique for recovering *Giardia* cysts and *Cryptosporidium* oocysts from fresh vegetables. Ann Ig 18:101–107

Di Benedetto MA, Cannova L, Di Piazza F, Amodio E, Bono F, Cerame G, Romano N (2007) Hygienic-sanitary quality of ready-to-eat salad vegetables on sale in the city of Palermo (Sicily). Ig Sanita Pubbl 63:659–670

Di Pinto A, Tantillo MG (2002) Direct detection of *Cryptosporidium parvum* oocysts by immuno-magnetic separation-polymerase chain reaction in raw milk. J Food Prot 65(8):1345–1348

Diallo MB, Anceno AJ, Tawatsupa B, Tripathi NK, Wangsuphachart V, Shipin OV (2009) GIS-based analysis of the fate of waste-related pathogens *Cryptosporidium parvum*, *Giardia lamblia* and *Escherichia coli* in a tropical canal network. J Water Health 7(1):133–143. doi:10.2166/wh.2009.010

Ditrich O, Palkovic L, Stěrba J, Prokopic J, Loudová J, Giboda M (1991) The first finding of *Cryptosporidium baileyi* in man. Parasitol Res 77(1):44–47

Dixon BR (2009) The role of livestock in the foodborne transmission of Giardia duodenalis and Cryptosporidium spp. to humans. In: Ortega-Pierres MG, Caccio SM, Fayer R, Smith H (eds) *Giardia* and *Cryptosporidium*: from molecules to disease. CAB International, Wallingford, pp 107–122

Dixon B, Parrington L, Cook A, Pollari F, Farber J (2013) Detection of *Cyclospora*, *Cryptosporidium*, and *Giardia* in ready-to-eat packaged leafy greens in Ontario, Canada. J Food Prot 76(2): 307–313. doi:10.4315/0362-028X.JFP-12-282

Downey AS, Graczyk TK (2007) Maximizing recovery and detection of *Cryptosporidium parvum* oocysts from spiked eastern oyster (*Crassostrea virginica*) tissue samples. Appl Environ Microbiol 73(21):6910–6915

Duffy G, McEvoy J, Moriarty EM, Sheridan JJ (2003) A study of *Cryptosporidium parvum* in beef. The National Food Centre, Research and Training for the Food Industry Research Report No. 62. Project RMIS No. 4723. ISBN 1 84170 337 0

Duhain GL, Minnaar A, Buys EM (2012) Effect of chlorine, blanching, freezing, and microwave heating on *Cryptosporidium parvum* viability inoculated on green peppers. J Food Prot 75(5): 936–941. doi:10.4315/0362-028X.JFP-11-367

DuPont HL, Chappell CL, Sterling CR, Okhuysen PC, Rose JB, Jakubowski W (1995) The infectivity of *Cryptosporidium parvum* in healthy volunteers. N Engl J Med 332(13):855–859

ECDC (2013) European Centre for Disease Prevention and Control. Cryptosporidium. Annual epidemiological report 2012. Reporting on 2010 surveillance data and 2011 epidemic intelligence data. ECDC, Stockholm. pp 72–74. http://www.ecdc.europa.eu/en/publications/Publications/Annual-Epidemiological-Report-2012.pdf

EFSA (2007) The community summary report on trends and sources of zoonoses, zoonotic agents, antimicrobial resistance and foodborne outbreaks in the European Union in 2006. EFSA J 130: 1–352

El Said Said D (2012) Detection of parasites in commonly consumed raw vegetables. Alexandria J Med 48:345–352

El Zawawy LA, El-Said D, Ali SM, Fathy FM (2010) Disinfection efficacy of sodium dichloroisocyanurate (NADCC) against common food-borne intestinal protozoa. J Egypt Soc Parasitol 40(1):165–185

Elwin K, Hadfield SJ, Robinson G, Chalmers RM (2012a) The epidemiology of sporadic human infections with unusual cryptosporidia detected during routine typing in England and Wales, 2000–2008. Epidemiol Infect 140(4):673–683

Elwin K, Hadfield SJ, Robinson G, Crouch ND, Chalmers RM (2012b) *Cryptosporidium viatorum* n. sp. (Apicomplexa: Cryptosporidiidae) among travellers returning to Great Britain from the Indian subcontinent, 2007–2011. Int J Parasitol 42(7):675–682

Ensink JH, van der Hoek W (2009) Implementation of the WHO guidelines for the safe use of wastewater in Pakistan: balancing risks and benefits. J Water Health 7(3):464–468. doi:10.2166/wh.2009.061

Ensink JH, Mahmood T, Dalsgaard A (2007) Wastewater-irrigated vegetables: market handling versus irrigation water quality. Trop Med Int Health 12(Suppl 2):2–7

Erickson MC, Ortega YR (2006) Inactivation of protozoan parasites in food, water, and environmental systems. J Food Prot 69(11):2786–2808

Ethelberg S, Lisby M, Vestergaard LS, Enemark HL, Olsen KE, Stensvold CR, Nielsen HV, Porsbo LJ, Plesner AM, Mølbak K (2009) A foodborne outbreak of *Cryptosporidium hominis* infection. Epidemiol Infect 137(3):348–356. doi:10.1017/S0950268808001817

Fayer R (2008) Biology. In: Faryer R, Xiao L (eds) *Cryptosporidium* and cryptosporidiosis, 2nd edn. CRC Press/IWA Publishing, Boca Raton, FL, pp 1–42

Fayer R, Graczyk TK, Lewis EJ, Trout JM, Farley CA (1998) Survival of infectious *Cryptosporidium parvum* oocysts in seawater and eastern oysters (*Crassostrea virginica*) in the Chesapeake Bay. Appl Environ Microbiol 64:1070–1074

Fayer R, Trout JM, Lewis EJ, Xiao L, Lal A, Jenkins MC, Graczyk TK (2002) Temporal variability of *Cryptosporidium* in the Chesapeake Bay. Parasitol Res 88:998–1003

Fayer R, Trout JM, Lewis EJ, Santin M, Zhou L, Lal AA, Xiao L (2003) Contamination of Atlantic coast commercial shellfish with *Cryptosporidium*. Parasitol Res 89:141–145

Fayer R, Santín M, Macarisin D (2010) *Cryptosporidium ubiquitum* n. sp. in animals and humans. Vet Parasitol 172(1–2):23–32

Fayer R, Santin M, Macarisin D, Bauchan G (2013) Adhesive-tape recovery combined with molecular and microscopic testing for the detection of *Cryptosporidium* oocysts on experimentally contaminated fresh produce and a food preparation surface. Parasitol Res 112(4): 1567–1574

Fetene T, Worku N, Huruy K, Kebede N (2011) *Cryptosporidium* recovered from *Musca domestica*, *Musca sorbens* and mango juice accessed by synanthropic flies in Bahirdar, Ethiopia. Zoonoses Public Health 58(1):69–75. doi:10.1111/j.1863-2378.2009.01298

Francavilla M, Trotta P, Marangi M, Breber P, Giangaspero A (2012) Environmental conditions in a lagoon and their possible effects on shellfish contamination by *Giardia* and *Cryptosporidium*. Aquac Int 20:707–724

Freidank H, Kist M (1987) Cryptosporidia in immunocompetent patients with gastroenteritis. Eur J Clin Microbiol 6(1):56–59

Freire-Santos F, Oteiza-Lopez AM, Vergara-Castiblanco CA, Ares-Mazas E, Alvarez-Suarez E, Garcia-Martin O (2000) Detection of *Cryptosporidium* oocysts in bivalve molluscs destined for human consumption. J Parasitol 86:853–854

Freire-Santos F, Oteiza-Lopez AM, Castro-Hermida JA, Garcia-Martin O, Ares-Mazas ME (2001) Viability and infectivity of oocysts recovered from clams, *Ruditapes philippinarum*, experimentally contaminated with *Cryptosporidium parvum*. Parasitol Res 87:428–430

Freire-Santos F, Gómez-Couso H, Ortega-Inarrea MR, Castro-Hermida JA, Oteiza-Lopez AM, Garcia-Martin O, Ares-Mazas ME (2002) Survival of *Cryptosporidium parvum* oocysts recovered from experimentally contaminated oysters (*Ostrea edulis*) and clams (*Tapes decussatus*). Parasitol Res 88:130–133

Freites A, Colmenares D, Pérez M, García M, Díaz de Suárez O (2009) *Cryptosporidium* sp infections and other intestinal parasites in food handlers from Zulia state, Venezuela. Invest Clin 50(1):13–21

Friedman DE, Patten KA, Rose JB, Barney MC (1997) The potential for *Cryptosporidium parvum* oocyst survival in beverages associated with contaminated tap water. J Food Safety 17: 125–132. doi:10.1111/j.1745-4565.1997.tb00181.x

Gale P (2005) Land application of treated sewage sludge: quantifying pathogen risks from consumption of crops. J Appl Microbiol 98(2):380–396

Garvey M, Farrell H, Cormican M, Rowan N (2010) Investigations of the relationship between use of *in vitro* cell culture-quantitative PCR and a mouse-based bioassay for evaluating critical factors affecting the disinfection performance of pulsed UV light for treating *Cryptosporidium parvum* oocysts in saline. J Microbiol Methods 80(3):267–273. doi:10.1016/j.mimet.2010.01.017

Gatei W, Suputtamongkol Y, Waywa D, Ashford RW, Bailey JW, Greensill J, Beeching NJ, Hart CA (2002) Zoonotic species of *Cryptosporidium* are as prevalent as the anthroponotic in HIV-infected patients in Thailand. Ann Trop Med Parasitol 96(8):797–802

Gatei W, Barrett D, Lindo JF, Eldemire-Shearer D, Cama V, Xiao L (2008) Unique *Cryptosporidium* population in HIV-infected persons, Jamaica. Emerg Infect Dis 14(5):841–843

Gelletlie R, Stuart J, Soltanpoor N, Armstrong R, Nichols G (1997) Cryptosporidiosis associated with school milk. Lancet 350(9083):1005–1006

Gherasim A, Lebbad M, Insulander M, Decraene V, Kling A, Hjertqvist M, Wallensten A (2012) Two geographically separated food-borne outbreaks in Sweden linked by an unusual *Cryptosporidium parvum* subtype, October 2010. Euro Surveill 17(46)

Ghimire TR, Mishra PN, Sherchand JB (2005) The seasonal outbreaks of *Cyclospora* and *Cryptosporidium* in Kathmandu, Nepal. J Nepal Health Res Counc 3:39–48. http://www.jnhrc. com.np/index.php?journal=nhrc&page=article&op=viewFile&path%5B%5D=77&path%5B %5D=120

Giangaspero A, Molini U, Iorio R, Traversa D, Paoletti B, Giansante C (2005) *Cryptosporidium parvum* oocysts in seawater clams (*Chamelea gallina*) in Italy. Prev Vet Med 69:203–212

Giangaspero A, Cirillo R, Lacasella V, Lonigro A, Marangi M, Cavallo P, Berrilli F, Di Cave D, Brandonisio O (2009) *Giardia* and *Cryptosporidium* in inflowing water and harvested shellfish in a lagoon in Southern Italy. Parasitol Int 58(1):12–17. doi:10.1016/j.parint.2008.07.003

Gil MI, Selma MV, López-Gálvez F, Allende A (2009) Fresh-cut product sanitation and wash water disinfection: problems and solutions. Int J Food Microbiol 134(1–2):37–45. doi:10.1016/j. ijfoodmicro.2009.05.021

Girotto KG, Grama DF, Cunha MJ, Faria ES, Limongi JE, Pinto Rde M, Cury MC (2013) Prevalence and risk factors for intestinal protozoa infection in elderly residents at Long Term Residency Institutions in Southeastern Brazil. Rev Inst Med Trop Sao Paulo 55(1):19–24

Gómez-Bautista M, Ortega-Mora LM, Tabares E, Lopez-Rodas V, Costas E (2000) Detection of infectious *Cryptosporidium parvum* oocysts in mussels (*Mytilus galloprovincialis*) and cockles (*Cerastoderma edule*). Appl Environ Microbiol 66:1866–1870

Gómez-Couso H, Freire-Santos F, Martinez-Urtaza J, Garcia-Martin O, Ares-Mazas ME (2003a) Contamination of bivalve molluscs by *Cryptosporidium* oocysts: the need for new quality control standards. Int J Food Microbiol 87:97–105

Gómez-Couso H, Freire-Santos F, Ortega-Inarrea MR, Castro-Hermida JA, Ares-Mazas ME (2003b) Environmental dispersal of *Cryptosporidium* parvum oocysts and cross transmission in cultured bivalve molluscs. Parasitol Res 90:140–142

Gómez-Couso H, Freire-Santos F, Amar CF, Grant KA, Williamson K, Ares-Mazas ME, McLauchlin J (2004) Detection of *Cryptosporidium* and *Giardia* in molluscan shellfish by multiplexed nested-PCR. Int J Food Microbiol 91:279–288

Gómez-Couso H, Mendez-Hermida F, Ares-Mazas E (2006a) Levels of detection of *Cryptosporidium* oocysts in mussels (*Mytilus galloprovincialis*) by IFA and PCR methods. Vet Parasitol 141: 60–65

Gómez-Couso H, Mendez-Hermida F, Castro-Hermida JA, Ares-Mazas E (2006b) Cooking mussels (*Mytilus galloprovincialis*) by steam does not destroy the infectivity of *Cryptosporidium parvum*. J Food Prot 69:948–950

Gómez-Couso H, Mendez-Hermida F, Castro-Hermida JA, Ares-Mazas E (2006c) *Cryptosporidium* contamination in harvesting areas of bivalve molluscs. J Food Prot 69:185–190

Goodgame RW, Genta RM, White AC, Chappell CL (1993) Intensity of infection in AIDS-associated cryptosporidiosis. J Infect Dis 167:704–709

Gortáres-Moroyoqui P, Castro-Espinoza L, Naranjo JE, Karpiscak MM, Freitas RJ, Gerba CP (2011) Microbiological water quality in a large irrigation system: El Valle del Yaqui, Sonora México. J Environ Sci Health A Tox Hazard Subst Environ Eng 46(14):1708–1712. doi:10.10 80/10934529.2011.623968

Grace D, Monda J, Karanja N, Randolph TF, Kang'ethe EK (2012) Participatory probabilistic assessment of the risk to human health associated with cryptosporidiosis from urban dairying in Dagoretti, Nairobi, Kenya. Trop Anim Health Prod 44(Suppl 1):S33–S40. doi:10.1007/s11250-012-0204-3

Gracenea M, Gómez MS, Ramírez CM (2011) Occurrence of *Cryptosporidium* oocysts and *Giardia* cysts in water from irrigation channels in Catalonia (NE Spain). Rev Ibero-Latinoam Parasitol 70:172–177

Graczyk TK, Conn DB, Marcogliese DJ, Graczyk H, De Lafontaine Y (2003) Accumulation of human waterborne parasites by zebra mussels (*Dreissena polymorpha*) and Asian freshwater clams (*Corbicula fluminea*). Parasitol Res 89(2):107–112

Graczyk TK, Girouard AS, Tamang L, Nappier SP, Schwab KJ (2006) Recovery, bioaccumulation, and inactivation of human waterborne pathogens by the Chesapeake Bay nonnative oyster, *Crassostrea ariakensis*. Appl Environ Microbiol 72:3390–3395

Graczyk TK, Lewis EJ, Glass G, Dasilva AJ, Tamang L, Girouard AS, Curriero FC (2007) Quantitative assessment of viable *Cryptosporidium parvum* load in commercial oysters (*Crassostrea virginica*) in the Chesapeake Bay. Parasitol Res 100:247–253

Greig JD, Todd EC, Bartleson CA, Michaels BS (2007) Outbreaks where food workers have been implicated in the spread of foodborne disease. Part 1. Description of the problem, methods, and agents involved. J Food Prot 70(7):1752–1761

Guerrant DI, Moore SR, Lima AA, Patrick PD, Schorling JB, Guerrant RL (1999) Association of early childhood diarrhea and cryptosporidiosis with impaired physical fitness and cognitive function four-seven years later in a poor urban community in northeast Brazil. Am J Trop Med Hyg 61(5):707–713

Hamilton AJ, Stagnitti F, Premier R, Boland A, Hale G (2006a) Quantitative microbial risk assessment modelling for the use of reclaimed water in irrigated horticulture. In: Brebbia CA, Popov V, Fayzieva D (eds) The sustainable world series, vol 13, Environmental health risk III. Wessex Institute of Technology, Ashurst, pp 71–81

Hamilton AJ, Stagnitti F, Premier R, Boland AM, Hale G (2006b) Quantitative microbial risk assessment models for consumption of raw vegetables irrigated with reclaimed water. Appl Environ Microbiol 72(5):3284–3290

Hanes DE, Worobo RW, Orlandi PA, Burr DH, Miliotis MD, Robl MG, Bier JW, Arrowood MJ, Churey JJ, Jackson GJ (2002) Inactivation of *Cryptosporidium parvum* oocysts in fresh apple cider by UV irradiation. Appl Environ Microbiol 68(8):4168–4172

Harp JA, Fayer R, Pesch BA, Jackson GJ (1996) Effect of pasteurization on infectivity of *Cryptosporidium parvum* oocysts in water and milk. Appl Environ Microbiol 62(8): 2866–2868

Harper CM, Cowell NA, Adams BC, Langley AJ, Wohlsen TD (2002) Outbreak of *Cryptosporidium* linked to drinking unpasteurised milk. Commun Dis Intell Q Rep 26(3):449–450

Hassan A, Farouk H, Hassanein F, Abdul-Ghani R (2011) Currency as a potential environmental vehicle for transmitting parasites among food-related workers in Alexandria, Egypt. Trans R Soc Trop Med Hyg 105(9):519–524. doi:10.1016/j.trstmh.2011.05.001

Hunter PR, Hughes S, Woodhouse S, Raj N, Syed Q, Chalmers RM, Verlander NQ, Goodacre J (2004a) Health sequelae of human cryptosporidiosis in immunocompetent patients. Clin Infect Dis 39:504–510

Hunter PR, Hughes S, Woodhouse S, Syed Q, Verlander NQ, Chalmers RM, Morgan K, Nichols G, Beeching N, Osborn K (2004b) Sporadic cryptosporidiosis case-control study with genotyping. Emerg Infect Dis 10(7):1241–1249

Hutchison ML, Walters LD, Moore T, Thomas DJ, Avery SM (2005) Fate of pathogens present in livestock wastes spread onto fescue plots. Appl Environ Microbiol 71(2):691–696

Insulander M, de Jong B, Svenungsson B (2008) A food-borne outbreak of cryptosporidiosis among guests and staff at a hotel restaurant in Stockholm county, Sweden, September 2008. Euro Surveill 13(51)

Insulander M, Silverlås C, Lebbad M, Karlsson L, Mattsson JG, Svenungsson B (2013) Molecular epidemiology and clinical manifestations of human cryptosporidiosis in Sweden. Epidemiol Infect 141(5):1009–1020

Izumi T, Yagita K, Endo T, Ohyama T (2006) Detection system of *Cryptosporidium parvum* oocysts by brackish water benthic shellfish (*Corbicula japonica*) as a biological indicator in river water. Arch Environ Contam Toxicol 51(4):559–566

Izumi T, Yagita K, Izumiyama S, Endo T, Itoh Y (2012) Depletion of *Cryptosporidium parvum* oocysts from contaminated sewage by using freshwater benthic pearl clams (*Hyriopsis schlegeli*). Appl Environ Microbiol 78(20):7420–7428

Jenkins M, Trout JM, Higgins J, Dorsch M, Veal D, Fayer R (2003) Comparison of tests for viable and infectious *Cryptosporidium parvum* oocysts. Parasitol Res 89(1):1–5

Jenkins MB, Eaglesham BS, Anthony LC, Kachlany SC, Bowman DD, Ghiorse WC (2010) Significance of wall structure, macromolecular composition, and surface polymers to the survival and transport of *Cryptosporidium parvum* oocysts. Appl Environ Microbiol 76(6): 1926–1934

Johannessen G, Robertson LJ, Myrmel M, Jensvoll L (2013) http://www.mattilsynet.no/mat_og_vann/uonskede_stofferimaten/smittestoffer_i_vegetabilske_naeringsmidler_2012.8961/BINARY/Smittestoffer%20i%20vegetabilske%20n%C3%A6ringsmidler%202012. Veterinærinstituttets rapportserie 7. ISSN: 1890-3290

Johnson AM, Giovanni GD, Rochelle PA (2012) Comparison of assays for sensitive and reproducible detection of cell culture-infectious *Cryptosporidium parvum* and *Cryptosporidium hominis* in drinking water. Appl Environ Microbiol 78(1):156–162. doi:10.1128/AEM.06444-11

Johnston SP, Ballard MM, Beach MJ, Causer L, Wilkins PP (2003) Evaluation of three commercial assays for detection of *Giardia* and *Cryptosporidium* organisms in fecal specimens. J Clin Microbiol 41:623–626

Kalisvaart BF (2004) Re-use of wastewater: preventing the recovery of pathogens by using medium-pressure UV lamp technology. Water Sci Technol 50(6):337–344

Karanis P, Kourenti C, Smith H (2007) Waterborne transmission of protozoan parasites: a world-wide review of outbreaks and lessons learnt. J Water Health 5:1–38

Keraita B, Konradsen F, Drechsel P, Abaidoo RC (2007) Effect of low-cost irrigation methods on microbial contamination of lettuce irrigated with untreated wastewater. Trop Med Int Health 12(Suppl 2):15–22

Keserue HA, Füchslin HP, Wittwer M, Nguyen-Viet H, Nguyen TT, Surinkul N, Koottatep T, Schürch N, Egli T (2012) Comparison of rapid methods for detection of *Giardia* spp. and *Cryptosporidium* spp. (oo)cysts using transportable instrumentation in a field deployment. Environ Sci Technol 46(16):8952–8959. doi:10.1021/es301974m

Khan SM, Debnath C, Pramanik AK, Xiao L, Nozaki T, Ganguly S (2010) Molecular characterization and assessment of zoonotic transmission of *Cryptosporidium* from dairy cattle in West Bengal, India. Vet Parasitol 171(1–2):41–47

King BJ, Monis PT (2007) Critical processes affecting *Cryptosporidium* oocyst survival in the environment. Parasitology 134(Pt 3):309–323

Kistemann T, Classen T, Koch C, Dangendorf F, Fischeder R, Gebel J, Vacata V, Exner M (2002) Microbial load of drinking water reservoir tributaries during extreme rainfall and runoff. Appl Environ Microbiol 68(5):2188–2197

Kniel KE, Lindsay DS, Sumner SS, Hackney CR, Pierson MD, Dubey JP (2002) Examination of attachment and survival of *Toxoplasma gondii* oocysts on raspberries and blueberries. J Parasitol 88(4):790–793

Kniel KE, Sumner SS, Lindsay DS, Hackney CR, Pierson MD, Zajac AM, Golden DA, Fayer R (2003) Effect of organic acids and hydrogen peroxide on *Cryptosporidium parvum* viability in fruit juices. J Food Prot 66(9):1650–1657

Kniel KE, Sumner SS, Pierson MD, Zajac AM, Hackney CR, Fayer R, Lindsay DS (2004) Effect of hydrogen peroxide and other protease inhibitors on *Cryptosporidium parvum* excystation and *in vitro* development. J Parasitol 90(4):885–888

Korich DG, Mead JR, Madore MS, Sinclair NA, Sterling CR (1990) Effects of ozone, chlorine dioxide, chlorine, and monochloramine on *Cryptosporidium parvum* oocyst viability. Appl Environ Microbiol 56(5):1423–1428

Kotloff KL, Nataro JP, Blackwelder WC, Nasrin D, Farag TH, Panchalingam S, Wu Y, Sow SO, Sur D, Breiman RF, Faruque AS, Zaidi AK, Saha D, Alonso PL, Tamboura B, Sanogo D, Onwuchekwa U, Manna B, Ramamurthy T, Kanungo S, Ochieng JB, Omore R, Oundo JO, Hossain A, Das SK, Ahmed S, Qureshi S, Quadri F, Adegbola RA, Antonio M, Hossain MJ, Akinsola A, Mandomando I, Nhampossa T, Acácio S, Biswas K, O'Reilly CE, Mintz ED, Berkeley LY, Muhsen K, Sommerfelt H, Robins-Browne RM, Levine MM (2013) Burden and aetiology of diarrhoeal disease in infants and young children in developing countries (the Global Enteric Multicenter Study, GEMS): a prospective, case-control study. Lancet 382(9888):209–222. doi:10.1016/S0140-6736(13)60844-2

Kumar P, Reinitz HW, Simunovic J, Sandeep KP, Franzon PD (2009) Overview of RFID technology and its applications in the food industry. J Food Sci 74(8):R101–R106. doi:10.1111/j.1750-3841.2009.01323.x

Kváč M, Kvetonová D, Sak B, Ditrich O (2009) *Cryptosporidium* pig genotype II in immunocompetent man. Emerg Infect Dis 15(6):982–983

Laberge I, Ibrahim A, Barta JR, Griffiths MW (1996) Detection of *Cryptosporidium parvum* in raw milk by PCR and oligonucleotide probe hybridization. Appl Environ Microbiol 62(9):3259–3264

Leal Diego AG, Pereira MA, Franco RMB, Branco N, Neto RC (2008) First report of *Cryptosporidium* spp. oocysts in oysters (*Crassostrea rhizophorae*) and cockles (*Tivela mactroides*) in Brazil. J Water Health 6:527–532

Leal Diego AG, Dores Ramos AP, Marques Souza DS, Durigan M, Greinert-Goulart JA, Moresco V, Amstutz RC, Micoli AH, Neto RC, Monte Barardi CR, Bueno Franco RM (2013) Sanitary quality of edible bivalve mollusks in Southeastern Brazil using an UV based depuration system. Ocean Coast Manag 72:93–100. doi:10.1016/j.ocecoaman.2011.07.010

Leander BS, Clopton R, Keeling P (2003) Phylogeny of gregarines (Apicomplexa) as inferred from small-subunit rDNA and beta-tubulin. Int J Syst Evol Microbiol 53:345–354

Leoni F, Gómez-Couso H, Ares-Mazás ME, McLauchlin J (2007) Multilocus genetic analysis of *Cryptosporidium* in naturally contaminated bivalve molluscs. J Appl Microbiol 103(6):2430–2437

Lévesque B, Gagnon F, Valentin A, Cartier JF, Chevalier P, Cardinal P, Cantin P, Gingras S (2006) A study to assess the microbial contamination of *Mya arenaria* clams from the north shore of the St Lawrence River estuary, (Québec, Canada). Can J Microbiol 52(10):984–991. doi:10.1139/w06-061

Lévesque B, Barthe C, Dixon BR, Parrington LJ, Martin D, Doidge B, Proulx JF, Murphy D (2010) Microbiological quality of blue mussels (*Mytilus edulis*) in Nunavik, Quebec: a pilot study. Can J Microbiol 56(11):968–977. doi:10.1139/w10-078

Li X, Guyot K, Dei-Cas E, Mallard JP, Ballet JJ, Brasseur P (2006) *Cryptosporidium* oocysts in mussels (*Mytilus edulis*) from Normandy (France). Int J Food Microbiol 108(3):321–325

Lonigro A, Pollice A, Spinelli R, Berrilli F, Di Cave D, D'Orazi C, Cavallo P, Brandonisio O (2006) *Giardia* cysts and *Cryptosporidium* oocysts in membrane-filtered municipal wastewater used for irrigation. Appl Environ Microbiol 72(12):7916–7918

Lorntz B, Soares AM, Moore SR, Pinkerton R, Gansneder B, Bovbjerg VE, Guyatt H, Lima AM, Guerrant RL (2006) Early childhood diarrhea predicts impaired school performance. Pediatr Infect Dis J 25(6):513–520

Lowery CJ, Nugent P, Moore JE, Millar BC, Xiru X, Dooley JS (2001) PCR-IMS detection and molecular typing of *Cryptosporidium parvum* recovered from a recreational river source and an associated mussel (*Mytilus edulis*) bed in Northern Ireland. Epidemiol Infect 127(3):545–553

Lucy FE, Graczyk TK, Tamang L, Miraflor A, Minchin D (2008) Biomonitoring of surface and coastal water for *Cryptosporidium*, *Giardia*, and human-virulent microsporidia using molluscan shellfish. Parasitol Res 103(6):1369–1375. doi:10.1007/s00436-008-1143-9

Mac Kenzie WR, Hoxie NJ, Proctor ME, Gradus MS, Blair KA, Peterson DE, Kazmierczak JJ, Addiss DG, Fox KR, Rose JB, Davis JP (1994) A massive outbreak in Milwaukee of *Cryptosporidium* infection transmitted through the public water supply. N Engl J Med 331(3):161–167

Macarisin D, Bauchan G, Fayer R (2010a) *Spinacia oleracea* L. leaf stomata harboring *Cryptosporidium parvum* oocysts: a potential threat to food safety. Appl Environ Microbiol 76(2):555–559. doi:10.1128/AEM.02118-09

Macarisin D, Santín M, Bauchan G, Fayer R (2010b) Infectivity of *Cryptosporidium parvum* oocysts after storage of experimentally contaminated apples. J Food Prot 73(10):1824–1829

Machado ECL, Stamford TLM, Alves LC, Melo RG, Shinohara NKS (2006) Effectiveness of *Cryptosporidium* spp. oocysts detection and enumeration methods in water and milk samples. Arq Bras Med Vet Zootec 58(3):432–439

MacRae M, Hamilton C, Strachan NJ, Wright S, Ogden ID (2005) The detection of *Cryptosporidium parvum* and *Escherichia coli* O157 in UK bivalve shellfish. J Microbiol Methods 60:395–401

Magkos F, Arvaniti F, Zampelas A (2006) Organic food: buying more safety or just peace of mind? A critical review of the literature. Crit Rev Food Sci Nutr 46(1):23–56

McDonald V (2008) Immune responses. In: Fayer R, Xiao L (eds) *Cryptosporidium* and Cryptosporidiosis, 2nd edn. CRC Press/IWA Publishing, Boca Raton, FL, pp 209–233

McEvoy JM, Moriarty EM, Duffy G, Sheridan JJ, Blair IS, McDowell DA (2004) Effect of a commercial freeze/tempering process on the viability of *Cryptosporidium parvum* oocysts on lean and fat beef trimmings. Meat Sci 67(4):559–564. doi:10.1016/j.meatsci.2003.12.007

Melo PC, Teodosio J, Reis J, Duarte A, Costa JC, Fonseca IP (2006) *Cryptosporidium* spp. in freshwater bivalves in Portugal. J Eukaryot Microbiol 53(Suppl 1):S28–S29

Millard PS, Gensheimer KF, Addiss DG, Sosin DM, Beckett GA, Houck-Jankoski A, Hudson A (1994) An outbreak of cryptosporidiosis from fresh-pressed apple cider. JAMA 272(20):1592–1596

Miller WA, Miller MA, Gardner IA, Atwill ER, Harris M, Ames J, Jessup D, Melli A, Paradies D, Worcester K, Olin P, Barnes N, Conrad PA (2005) New genotypes and factors associated with *Cryptosporidium* detection in mussels (*Mytilus* spp.) along the California coast. Int J Parasitol 35:1103–1113

Miller WA, Gardner IA, Atwill ER, Leutenegger CM, Miller MA, Hedrick RP, Melli AC, Barnes NM, Conrad PA (2006) Evaluation of methods for improved detection of *Cryptosporidium* spp. in mussels (*Mytilus californianus*). J Microbiol Methods 65:367–379

Minarovicová J, Kaclikova E, Krascsenicsova K, Siekel P (2007) Detection of *Cryptosporidium parvum* by polymerase chain reaction. J Food Nutr Res 46(2):58–62

Minarovicova J, Lopasovska J, Valik L, Kuchta T (2011) A method for the detection of *Cryptosporidium parvum* oocysts in milk based on microfiltration and real-time polymerase chain reaction. Food Anal Methods 4(1):116–120

Miraglia M, Marvin HJ, Kleter GA, Battilani P, Brera C, Coni E, Cubadda F, Croci L, De Santis B, Dekkers S, Filippi L, Hutjes RW, Noordam MY, Pisante M, Piva G, Prandini A, Toti L, van den Born GJ, Vespermann A (2009) Climate change and food safety: an emerging issue with special focus on Europe. Food Chem Toxicol 247(5):1009–1021. doi:10.1016/j.fct.2009.02.005

Mladineo I, Trumbić Z, Jozić S, Šegvić T (2009) First report of *Cryptosporidium* sp. (Coccidia, Apicomplexa) oocysts in the black mussel (*Mytilus galloprovincialis*) reared in the Mali Ston Bay, Adriatic Sea. J Shellfish Res 28:541–546

Molini U, Traversa D, Ceschia G, Iorio R, Boffo L, Zentilin A, Capelli G, Giangaspero A (2007) Temporal occurrence of *Cryptosporidium* in the Manila clam *Ruditapes philippinarum* in northern Adriatic Italian lagoons. J Food Prot 70(2):494–499

Moore JE, Millar BC, Kenny F, Lowery CJ, Xiao L, Rao JR, Nicholson V, Watabe M, Heaney N, Sunnotel O, McCorry K, Rooney PJ, Snelling WJ, Dooley JSG (2007) Detection of *Cryptosporidium parvum* in lettuce. Int J Food Sci Technol 42:385–393. doi:10.1111/j.1365-2621.2006.01235.x

Moriarty EM, McEvoy JM, Duffy G, Sheridan JJ, Blair IS, McDowell DA (2004) Development of a novel method for isolating and detecting *Cryptosporidium parvum* from lean and fat beef carcass surfaces. Food Microbiol 21:275–282

Moriarty EM, Duffy G, McEvoy JM, Caccio S, Sheridan JJ, McDowell D, Blair IS (2005) The effect of thermal treatments on the viability and infectivity of *Cryptosporidium parvum* on beef surfaces. J Appl Microbiol 98(3):618–623

Morrison DA (2009) Evolution of the Apicomplexa: where are we now? Trends Parasitol 25(8): 375–382

Morse TD, Nichols RA, Grimason AM, Campbell BM, Tembo KC, Smith HV (2007) Incidence of cryptosporidiosis species in paediatric patients in Malawi. Epidemiol Infect 135(8):1307–1315

Mosier DA, Oberst RD (2000) Cryptosporidiosis. A global challenge. Ann N Y Acad Sci 916:102–111

Mossallam SF (2010) Detection of some intestinal protozoa in commercial fresh juices. J Egypt Soc Parasitol 40(1):135–149

Mota A, Mena KD, Soto-Beltran M, Tarwater PM, Cháidez C (2009) Risk assessment of *Cryptosporidium* and *Giardia* in water irrigating fresh produce in Mexico. J Food Prot 72(10): 2184–2188

Muthusamy D, Rao SS, Ramani S, Monica B, Banerjee I, Abraham OC, Mathai DC, Primrose B, Muliyil J, Wanke CA, Ward HD, Kang G (2006) Multilocus genotyping of *Cryptosporidium* sp. isolates from human immunodeficiency virus-infected individuals in South India. J Clin Microbiol 44(2):632–634

Nappier SP, Graczyk TK, Tamang L, Schwab KJ (2010) Co-localized *Crassostrea virginica* and *Crassostrea ariakensis* oysters differ in bioaccumulation, retention and depuration of microbial indicators and human enteropathogens. J Appl Microbiol 108(2):736–744. doi:10.1111/j.1365-2672.2009.04480.x

Negm AY (2003) Human pathogenic protozoa in bivalves collected from local markets in Alexandria. J Egypt Soc Parasitol 33:991–998

Ng-Hublin JS, Combs B, Mackenzie B, Ryan U (2013) Human cryptosporidiosis diagnosed in Western Australia: a mixed infection with *Cryptosporidium meleagridis*, the *Cryptosporidium* mink genotype, and an unknown *Cryptosporidium* species. J Clin Microbiol 51(7):2463–2465

Niehaus MD, Moore SR, Patrick PD, Derr LL, Lorntz B, Lima AA, Guerrant RL (2002) Early childhood diarrhea is associated with diminished cognitive function 4 to 7 years later in children in a northeast Brazilian shantytown. Am J Trop Med Hyg 66(5):590–593

Nuñez FA, Robertson LJ (2012) Trends in foodborne diseases, including deaths. In: Robertson LJ, Smith HV (eds) Foodborne protozoan parasite. Nova Publishers, Hauppauge, NY

Nydam DV, Wade SE, Schaaf SL, Mohammed HO (2001) Number of *Cryptosporidium parvum* oocysts or *Giardia* spp cysts shed by dairy calves after natural infection. Am J Vet Res 62(10): 1612–1615

Okhuysen PC, Chappell CL, Crabb JH, Sterling CR, DuPont HL (1999) Virulence of three distinct *Cryptosporidium parvum* isolates for healthy adults. J Infect Dis 180(4):1275–1281

Okhuysen PC, Rich SM, Chappell CL, Grimes KA, Widmer G, Feng X, Tzipori S (2002) Infectivity of a *Cryptosporidium parvum* isolate of cervine origin for healthy adults and interferon-gamma knockout mice. J Infect Dis 185(9):1320–1325

Orta de Velásquez MT, Rojas-Valencia MN, Reales-Pineda AC (2006) Evaluation of phytotoxic elements, trace elements and nutrients in a standardized crop plant, irrigated with raw wastewater treated by APT and ozone. Water Sci Technol 54(11–12):165–173

Ortega YR, Liao J (2006) Microwave inactivation of *Cyclospora cayetanensis* sporulation and viability of *Cryptosporidium parvum* oocysts. J Food Prot 69(8):1957–1960

Ortega YR, Roxas CR, Gilman RH, Miller NJ, Cabrera L, Taquiri C, Sterling CR (1997) Isolation of *Cryptosporidium parvum* and *Cyclospora cayetanensis* from vegetables collected in markets of an endemic region in Peru. Am J Trop Med Hyg 57(6):683–686

Ortega YR, Mann A, Torres MP, Cama V (2008) Efficacy of gaseous chlorine dioxide as a sanitizer against *Cryptosporidium parvum*, *Cyclospora cayetanensis*, and *Encephalitozoon intestinalis* on produce. J Food Prot 71(12):2410–2414

Ortega YR, Torres MP, Tatum JM (2011) Efficacy of levulinic acid-sodium dodecyl sulfate against *Encephalitozoon intestinalis*, *Escherichia coli* O157:H7, and *Cryptosporidium parvum*. J Food Prot 74(1):140–144. doi:10.4315/0362-028X.JFP-10-104

Palmer CJ, Xiao L, Terashima A, Guerra H, Gotuzzo E, Saldías G, Bonilla JA, Zhou L, Lindquist A, Upton SJ (2003) *Cryptosporidium muris*, a rodent pathogen, recovered from a human in Perú. Emerg Infect Dis 9(9):1174–1176

Paruch AM, Mæhlum T, Robertson LJ (submitted) Changes in microbial quality of irrigation water under different weather conditions in Southeast Norway. J Water Clim Change

Pönka A, Kotilainen H, Rimhanen-Finne R, Hokkanen P, Hänninen ML, Kaarna A, Meri T, Kuusi M (2009) A foodborne outbreak due to *Cryptosporidium parvum* in Helsinki, November 2008. Euro Surveill 14(28)

Quiroz ES, Bern C, MacArthur JR, Xiao L, Fletcher M, Arrowood MJ, Shay DK, Levy ME, Glass RI, Lal A (2000) An outbreak of cryptosporidiosis linked to a foodhandler. J Infect Dis 181(2):695–700

Raccurt CP (2007) Worldwide human zoonotic cryptosporidiosis caused by *Cryptosporidium felis*. Parasite 14(1):15–20

Rai AK, Chakravorty R, Paul J (2008) Detection of *Giardia*, *Entamoeba*, and *Cryptosporidium* in unprocessed food items from northern India. World J Microbiol Biotechnol 24(12): 2879–2887

Ranjbar-Bahadori S, Mostoophi A, Shemshadi B (2013) Study on *Cryptosporidium* contamination in vegetable farms around Tehran. Trop Biomed 30(2):193–198

Rasková V, Kvetonová D, Sak B, McEvoy J, Edwinson A, Stenger B, Kváč M (2013) Human cryptosporidiosis caused by *Cryptosporidium tyzzeri* and *C. parvum* isolates presumably transmitted from wild mice. J Clin Microbiol 51(1):360–362

Rimšeliené G, Vold L, Robertson L, Nelke C, Søli K, Johansen ØH, Thrana FS, Nygård K (2011) An outbreak of gastroenteritis among schoolchildren staying in a wildlife reserve: thorough investigation reveals Norway's largest cryptosporidiosis outbreak. Scand J Public Health 39: 287–295

Ripabelli G, Leone A, Sammarco ML, Fanelli I, Grasso GM, McLauchlin J (2004) Detection of *Cryptosporidium parvum* oocysts in experimentally contaminated lettuce using filtration, immunomagnetic separation, light microscopy, and PCR. Foodborne Pathog Dis 1:216–222

Robertson LJ (2007) The potential for marine bivalve shellfish to act as transmission vehicles for outbreaks of protozoan infections in humans: a review. Int J Food Microbiol 120(3):201–216

Robertson LJ (2013) *Giardia* as a foodborne pathogen, Springer briefs in food, health and nutrition. Springer, New York. ISBN 978-1461477556

Robertson LJ (2014) Chapter 4.2. Waterborne Zoonoses—*Cryptosporidium* and cryptosporidiosis: a small parasite that makes a big splash. In: Sing A (ed). Zoonoses: infections affecting humans and animals—a focus on public health aspects. Springer

Robertson LJ, Sprong H, Ortega Y, van der Giessen JWB, Fayer R (2014) Impacts of globalisation on foodborne parasites. Journal Trends Parasitol (in press)

Robertson LJ, Chalmers RM (2013) Foodborne cryptosporidiosis: is there really more in Nordic countries? Trends Parasitol 29(1):3–9. doi:10.1016/j.pt.2012.10.003

Robertson LJ, Fayer R (2012) Cryptosporidium. In: Robertson LJ, Smith HV (eds) Foodborne protozoan parasite. Nova Publishers, Hauppauge, NY

Robertson LJ, Gjerde B (2000) Isolation and enumeration of *Giardia* cysts, *Cryptosporidium* oocysts, and *Ascaris* eggs from fruits and vegetables. J Food Prot 63:775–778

Robertson LJ, Gjerde B (2001a) Factors affecting recovery efficiency in isolation of *Cryptosporidium* oocysts and *Giardia* cysts from vegetables for standard method development. J Food Prot 64(11):1799–1805

Robertson LJ, Gjerde BK (2001b) Occurrence of parasites on fruits and vegetables in Norway. J Food Prot 64:1793–1798

Robertson LJ, Gjerde BK (2006) Fate of *Cryptosporidium* oocysts and *Giardia* cysts in the Norwegian aquatic environment over winter. Microb Ecol 52:597–602. doi:10.1007/s00248-006-9005-4

Robertson LJ, Gjerde BK (2007) *Cryptosporidium* oocysts: challenging adversaries? Trends Parasitol 23(8):344–347

Robertson LJ, Gjerde B (2008) Development and use of a pepsin digestion method for analysis of shellfish for *Cryptosporidium* oocysts and *Giardia* cysts. J Food Prot 71(5):959–966

Robertson LJ, Huang Q (2012) Analysis of cured meat products for *Cryptosporidium* oocysts following possible contamination during an extensive waterborne outbreak of cryptosporidiosis. J Food Prot 75(5):982–988. doi:10.4315/0362-028X.JFP-11-525

Robertson LJ, Campbell AT, Smith HV (1992) Survival of *Cryptosporidium parvum* oocysts under various environmental pressures. Appl Environ Microbiol 58(11):3494–3500

Robertson B, Sinclair MI, Forbes AB, Veitch M, Kirk M, Cunliffe D, Willis J, Fairley CK (2002) Case-control studies of sporadic cryptosporidiosis in Melbourne and Adelaide, Australia. Epidemiol Infect 128:419–431

Robertson LJ, Forberg T, Hermansen L, Gjerde BK, Alvsvåg JO, Langeland N (2006) *Cryptosporidium parvum* infections in Bergen, Norway, during an extensive outbreak of waterborne giardiasis in autumn and winter 2004. Appl Environ Microbiol 72(3):2218–2220

Robertson LJ, van der Giessen JWB, Batz MB, Kojima M, Cahill S (2013) Have foodborne parasites finally become a global concern? Trends Parasitol 29(3):101–103. doi:10.1016/j.pt.2012.12.004

Robinson G, Chalmers RM (2010) The European rabbit (*Oryctolagus cuniculus*), a source of zoonotic cryptosporidiosis. Zoonoses Public Health 57(7–8):e1–e13

Robinson G, Chalmers RM (2012) Assessment of polymorphic genetic markers for multi-locus typing of *Cryptosporidium parvum* and *Cryptosporidium hominis*. Exp Parasitol 132(2):200–215

Rochelle PA, Marshall MM, Mead JR, Johnson AM, Korich DG, Rosen JS, De Leon R (2002) Comparison of *in vitro* cell culture and a mouse assay for measuring infectivity of *Cryptosporidium parvum*. Appl Environ Microbiol 68(8):3809–3817

Rodríguez DC, Pino N, Peñuela G (2012) Microbiological quality indicators in waters of dairy farms: detection of pathogens by PCR in real time. Sci Total Environ 427–428:314–318. doi:10.1016/j.scitotenv.2012.03.052

Romanova TV, Shkarin VV, Khazenson LB (1992) Group cryptosporidiosis morbidity in children. Med Parazitol (Mosk) (3):50–52

Rosenblum J, Ge C, Bohrerova Z, Yousef A, Lee J (2012) Ozonation as a clean technology for fresh produce industry and environment: sanitizer efficiency and wastewater quality. J Appl Microbiol 113(4):837–845. doi:10.1111/j.1365-2672.2012.05393.x

Roy SL, DeLong SM, Stenzel SA, Shiferaw B, Roberts JM, Khalakdina A, Marcus R, Segler SD, Shah DD, Thomas S, Vugia DJ, Zansky SM, Dietz V, Beach MJ, Emerging Infections Program FoodNet Working Group (2004) Risk factors for sporadic cryptosporidiosis among immunocompetent persons in the United States from 1999 to 2001. J Clin Microbiol 42(7):2944–2951

Ryan U, Power M (2012) *Cryptosporidium* species in Australian wildlife and domestic animals. Parasitology 139(13):1673–1688

Rzeżutka A, Nichols RA, Connelly L, Kaupke A, Kozyra I, Cook N, Birrell S, Smith HV (2010) *Cryptosporidium* oocysts on fresh produce from areas of high livestock production in Poland. Int J Food Microbiol 139:96–101

Savioli L, Smith H, Thompson A (2006) *Giardia* and *Cryptosporidium* join the 'Neglected Diseases Initiative'. Trends Parasitol 22(5):203–208

Scallan E, Hoekstra RM, Angulo FJ, Tauxe RV, Widdowson MA, Roy SL, Jones JL, Griffin PM (2011) Foodborne illness acquired in the United States–major pathogens. Emerg Infect Dis 17:7–15. doi:10.3201/eid1701.091101p1

Schets FM, van den Berg HH, Engels GB, Lodder WJ, de Roda Husman AM (2007) *Cryptosporidium* and *Giardia* in commercial and non-commercial oysters (*Crassostrea gigas*) and water from the Oosterschelde, The Netherlands. Int J Food Microbiol 113:189–194

Schets FM, van den Berg HH, de Roda Husman AM (2013) Determination of the recovery efficiency of *Cryptosporidium* oocysts and *Giardia* cysts from seeded bivalve mollusks. J Food Prot 76(1):93–98

Schijven JF, Teunis PF, Rutjes SA, Bouwknegt M, de Roda Husman AM (2011) QMRAspot: a tool for Quantitative Microbial Risk Assessment from surface water to potable water. Water Res 45(17):5564–5576. doi:10.1016/j.watres.2011.08.024

Schijven J, Bouwknegt M, de Roda Husman AM, Rutjes S, Sudre B, Suk JE, Semenza JC (2013) A decision support tool to compare waterborne and foodborne infection and/or illness risks associated with climate change. Risk Anal. In press. doi: 10.1111/risa.12077

Selma MV, Allende A, López-Gálvez F, Conesa MA, Gil MI (2008a) Heterogeneous photocatalytic disinfection of wash waters from the fresh-cut vegetable industry. J Food Prot 71(2): 286–292

Selma MV, Allende A, López-Gálvez F, Conesa MA, Gil MI (2008b) Disinfection potential of ozone, ultraviolet-C and their combination in wash water for the fresh-cut vegetable industry. Food Microbiol 25(6):809–814. doi:10.1016/j.fm.2008.04.005

Shapiro K, Miller WA, Silver MW, Odagiri M, Largier JL, Conrad PA, Mazet JA (2013) Research commentary: association of zoonotic pathogens with fresh, estuarine, and marine macroaggregates. Microb Ecol 65(4):928–933. doi:10.1007/s00248-012-0147-2

Shields JM, Joo J, Kim R, Murphy HR (2013) Assessment of three commercial DNA extraction kits and a laboratory-developed method for detecting *Cryptosporidium* and *Cyclospora* in raspberry wash, basil wash and pesto. J Microbiol Methods 92(1):51–58

Shirley DA, Moonah SN, Kotloff KL (2012) Burden of disease from cryptosporidiosis. Curr Opin Infect Dis 25(5):555–563

Shuval H, Lampert Y, Fattal B (1997) Development of a risk assessment approach for evaluating wastewater reuse standards for agriculture. Water Sci Technol 35(11–12):15–20. doi:10.1016/S0273-1223(97)00228-X

Silverlås C, Mattsson JG, Insulander M, Lebbad M (2012) Zoonotic transmission of *Cryptosporidium meleagridis* on an organic Swedish farm. Int J Parasitol 42(11):963–967

Slifko TR, Raghubeer E, Rose JB (2000) Effect of high hydrostatic pressure on *Cryptosporidium parvum* infectivity. J Food Prot 63(9):1262–1267

Smeets PW, Rietveld LC, van Dijk JC, Medema GJ (2010) Practical applications of quantitative microbial risk assessment (QMRA) for water safety plans. Water Sci Technol 61(6):1561–1568. doi:10.2166/wst.2010.839

Souza DS, Ramos AP, Nunes FF, Moresco V, Taniguchi S, Leal DA, Sasaki ST, Bícego MC, Montone RC, Durigan M, Teixeira AL, Pilotto MR, Delfino N, Franco RM, Melo CM, Bainy AC, Barardi CR (2012) Evaluation of tropical water sources and mollusks in southern Brazil using microbiological, biochemical, and chemical parameters. Ecotoxicol Environ Saf 76(2): 153–161. doi:10.1016/j.ecoenv.2011.09.018

Srikanth R, Naik D (2004) Health effects of wastewater reuse for agriculture in the suburbs of Asmara city, Eritrea. Int J Occup Environ Health 10(3):284–288

Srisuphanunt M, Wiwanitkit V, Saksirisampant W, Karanis P (2009) Detection of *Cryptosporidium* oocysts in green mussels (*Perna viridis*) from shell-fish markets of Thailand. Parasite 16(3): 235–239

Stark D, Al-Qassab SE, Barratt JL, Stanley K, Roberts T, Marriott D, Harkness J, Ellis JT (2011) Evaluation of multiplex tandem real-time PCR for detection of *Cryptosporidium* spp., *Dientamoeba fragilis*, *Entamoeba histolytica*, and *Giardia intestinalis* in clinical stool samples. J Clin Microbiol 49(1):257–262. doi:10.1128/JCM.01796-10

Stensvold CR, Lebbad M, Verweij JJ (2011) The impact of genetic diversity in protozoa on molecular diagnostics. Trends Parasitol 27(2):53–58. doi:10.1016/j.pt.2010.11.005

Sulaiman IM, Xiao L, Yang C, Escalante L, Moore A, Beard CB, Arrowood MJ, Lal AA (1998) Differentiating human from animal isolates of *Cryptosporidium parvum*. Emerg Infect Dis 4: 681–685

Sulaiman IM, Hira PR, Zhou L, Al-Ali FM, Al-Shelahi FA, Shweiki HM, Iqbal J, Khalid N, Xiao L (2005) Unique endemicity of cryptosporidiosis in children in Kuwait. J Clin Microbiol 43(6):2805–2809

Sunnotel O, Snelling WJ, McDonough N, Browne L, Moore JE, Dooley JS, Lowery CJ (2007) Effectiveness of standard UV depuration at inactivating *Cryptosporidium parvum* recovered from spiked Pacific oysters (*Crassostrea gigas*). Appl Environ Microbiol 73(16):5083–5087

Sutthikornchai C, Jantanavivat C, Thongrungkiat S, Harnroongroj T, Sukthana Y (2005) Protozoal contamination of water used in Thai frozen food industry. Southeast Asian J Trop Med Public Health 36(Suppl 4):41–45

Tamburrini A, Pozio E (1999) Long-term survival of *Cryptosporidium parvum* oocysts in seawater and in experimentally infected mussels (*Mytilus galloprovincialis*). Int J Parasitol 29:711–715

Tang J, McDonald S, Peng X, Samadder SR, Murphy TM, Holden NM (2011) Modelling *Cryptosporidium* oocysts transport in small ungauged agricultural catchments. Water Res 45(12):3665–3680. doi:10.1016/j.watres.2011.04.013

Taniuchi M, Verweij JJ, Noor Z, Sobuz SU, Lieshout L, Petri WA Jr, Haque R, Houpt ER (2011) High throughput multiplex PCR and probe-based detection with Luminex beads for seven intestinal parasites. Am J Trop Med Hyg 84(2):332–337. doi:10.4269/ajtmh.2011.10-0461

Tedde T, Piras G, Salza S, Nives RM, Sanna G, Tola S, Culurgioni J, Piras C, Merella P, Garippa G, Virgilio S (2013) Investigation into *Cryptosporidium* and *Giardia* in bivalve mollusks farmed in Sardinia region and destined for human consumption. Italian J Food Safety 2:e26. doi:10.4081/ijfs.2013.e26

Teunis PF, Chappell CL, Okhuysen PC (2002a) *Cryptosporidium* dose response studies: variation between isolates. Risk Anal 22:175–183

Teunis PF, Chappell CL, Okhuysen PC (2002b) *Cryptosporidium* dose-response studies: variation between hosts. Risk Anal 22:475–485

Thurston-Enriquez JA, Watt P, Dowd SE, Enriquez R, Pepper IL, Gerba CP (2002) Detection of protozoan parasites and microsporidia in irrigation waters used for crop production. J Food Prot 65:378–382

Traversa D, Giangaspero A, Molini U, Iorio R, Paoletti B, Otranto D, Giansante C (2004) Genotyping of *Cryptosporidium* isolates from *Chamelea gallina* clams in Italy. Appl Environ Microbiol 70:4367–4370

Uga S, Matsuo J, Kono E, Kimura K, Inoue M, Rai SK, Ono K (2000) Prevalence of *Cryptosporidium parvum* infection and pattern of oocyst shedding in calves in Japan. Vet Parasitol 94:27–32

Unnevehr LJ (2000) Food safety issues and fresh food product exports from LDCs. Agric Econ 23: 231–240

Vuong TA, Nguyen TT, Klank LT, Phung DC, Dalsgaard A (2007) Faecal and protozoan parasite contamination of water spinach (*Ipomoea aquatica*) cultivated in urban wastewater in Phnom Penh, Cambodia. Trop Med Int Health 12:73–81

Wang L, Zhang H, Zhao X, Zhang L, Zhang G, Guo M, Liu L, Feng Y, Xiao L (2013) Zoonotic *Cryptosporidium* species and *Enterocytozoon bieneusi* genotypes in HIV-positive patients on antiretroviral therapy. J Clin Microbiol 51(2):557–563

Warnecke M, Weir C, Vesey G (2003) Evaluation of an internal positive control for *Cryptosporidium* and *Giardia* testing in water samples. Lett Appl Microbiol 37(3):244–248

Weitzel T, Dittrich S, Möhl I, Adusu E, Jelinek T (2006) Evaluation of seven commercial antigen detection tests for *Giardia* and *Cryptosporidium* in stool samples. Clin Microbiol Infect 12(7):656–659

Westrell T, Schönning C, Stenström TA, Ashbolt NJ (2004) QMRA (quantitative microbial risk assessment) and HACCP (hazard analysis and critical control points) for management of pathogens in wastewater and sewage sludge treatment and reuse. Water Sci Technol 50(2): 23–30

Willis JE, Greenwood S, Spears J, Davidson J, McClure C, McClure JT (2012). Ability of oysters (*Crassostrea virginica*) to harbour zoonotic parasites *Cryptosporidium parvum* and *Giardia duodenalis* during constant or limited exposures in a static tank system. Oral Presentation, IV International *Giardia & Cryptosporidum* Conference, Wellington, New Zealand. Page 126 of Abstract book

Willis JE, McClure JT, Davidson J, McClure C, Greenwood SJ (2013) Global occurrence of *Cryptosporidium* and *Giardia* in shellfish: should Canada take a closer look? Food Res Int 52(1):119–135

Wolfenden L, Wyse RJ, Britton BI, Campbell KJ, Hodder RK, Stacey FG, McElduff P, James EL (2012) Interventions for increasing fruit and vegetable consumption in children aged 5 years and under. Cochrane Database Syst Rev 11, CD008552. doi:10.1002/14651858.CD008552.pub2

Xiao L (2010) Molecular epidemiology of cryptosporidiosis: an update. Exp Parasitol 124(1): 80–89

Yang R, Murphy C, Song Y, Ng-Hublin J, Estcourt A, Hijjawi N, Chalmers R, Hadfield S, Bath A, Gordon C, Ryan U (2013). Specific and quantitative detection and identification of *Cryptosporidium hominis* and *C. parvum* in clinical and environmental samples. Exp Parasitol 135(1):142–147.

Yoshida H, Matsuo M, Miyoshi T, Uchino K, Nakaguchi H, Fukumoto T, Teranaka Y, Tanaka T (2007) An outbreak of cryptosporidiosis suspected to be related to contaminated food, October 2006, Sakai City, Japan. Jpn J Infect Dis 60(6):405–407

Zhao T, Zhao P, Doyle MP (2009) Inactivation of *Salmonella* and *Escherichia coli* O157:H7 on lettuce and poultry skin by combinations of levulinic acid and sodium dodecyl sulfate. J Food Prot 72(5):928–936